P9-DCY-344

NORTHAMPTON COMMUNITY
DISCARDED
COLLEGE LIBRARY

A Clone of Your Own?

"a wonderfully accessible account of the science involved in cloning and of the moral issues that surround it"

Mary Warnock

Seeing double

REDIFFUSION

A Clone of Your Own?

The Science and Ethics of Cloning

Arlene Judith Klotzko

With original drawings by David Mann

CAMBRIDGE UNIVERSITY PRESS
Cambridge, New York, Melbourne, Madrid, Cape Town, Singapore, Sào Paulo

Cambridge University Press
40 West 20th Street, New York, NY 10011–4211, USA

Published in the United States of America by Cambridge University Press, New York

www.cambridge.org
Information on this title:www.cambridge.org/9780521460736

First published 2006

Printed in the United States of America

A catalogue record for this book is available from the British Library

ISBN-13 978-0-521-85294-4 hardback
ISBN-10 0-521-85294-3 hardback

In loving memory of my father, Charles Klotzko, who taught me, by his example, to notice—and to cherish—what is unique in everyone.

Acknowledgements

A generous grant from the Alfred P. Sloan Foundation supported the research and writing of this book. I would like to thank the Foundation and, in particular, its programme officer for the Public Understanding of Science, Doron Weber.

David Mann provided marvellous drawings that enriched and enlivened the text.

Many friends and colleagues have also contributed in various ways and I am grateful to all of them: Carol Baron, Colin Blakemore, Gregory Bock, Kenneth Boyd, Keith Campbell, Elizabeth Graham, Julia Greenstein, Christopher Higgins, Susan Joffe, Bob Lieberman, Anne McLaren, Laurie Petrick, Tony Rawsthorne, Martin Redfern, Jon Risley, Eric Sharps, Virginia Schultz, Azim Surani, Alan Trounson, Robin Weiss, and Steen Willadsen.

Dr Lindsay Sharp and Heather Mayfield, respectively the Director of the National Museum of Science and Industry and Deputy Director of the Science Museum, have provided me with the wonderful opportunity to be the Museum's Writer in Residence.

My agents, Daniela Bernardelle and Bruce Hunter of David Higham Associates, have been extraordinarily helpful at every stage.

Special thanks to Shelley Cox who shared and helped to shape my vision for this book. Thanks also to Emma Simmons, Kirk Jensen, Michael Rodgers, Marsha Filion, Mary Worthington, and an anonymous reader who made suggestions that improved the final product.

Contents

1. The Cholmondeley Sisters

List of Illustrations

Original drawings by David Mann

2. Making faces

Introduction: Facts and Fictions

When George Lucas sat down to choose the title of his much-anticipated *Star Wars* 'prequel', he came up with *Attack of the Clones*. Indeed, it would have been hard to find a title with associations that were more negative, more menacing, and more calculated to induce a visceral shudder. Just the use of the word 'clone' gives us a rich store of visual images on which to draw. Science fiction films had set the pattern; nameless, faceless, mindless pseudo-humans march across our mental landscape. In film, the dream of cloning is almost always a nightmare.

Powerful images and metaphors of myth, literature, and film that invade our thoughts and colour our responses to cloning retain their hold on our imaginations because they address age-old questions and concerns. Concerns about the hubris-afflicted or simply mad scientist—Frankenstein is, of course, the paradigm—creating life, violating the natural order, and unleashing forces beyond his or our control. And concerns about the dehumanization of the industrial age so memorably depicted in Huxley's *Brave New World*.

As we drink our morning coffee and read the paper, we find that the nightmare has invaded our kitchen in the guise of an odd cast of characters—a quartet of would-be human

3. Space invaders

cloners. Richard Seed, a physicist, who has about as much chance of actually cloning anyone as I do, was the first to appear on the scene. He promised that cloning could put us at one with God, and he volunteered himself along with his wife's womb for that holy purpose. Then came Rael, a former journalist, now leader of a religious cult. In late December 2002 and early January 2003, his human cloning company, Clonaid, became notorious and then infamous for its unsubstantiated claims to have cloned babies.

Rael maintains that he met clones when their UFO landed in Canada. It was then that he got his cloning instructions.

Among Rael's bizarre and preposterous promises is the resurrection by cloning of all those who died in the September 11th attack on the United States. The victims, he says, would be restored to their grieving loved ones and the hijacker clones could be put on trial, thereby giving us justice and mercy all in one neat scientific package.

The third member of the cloning quartet is the rather operatic Italian fertility specialist, Severino Antinori, until recently best known for using IVF techniques to give the world its first 63-year-old mother. His promises, boasts, and frequent updates on the progress of his efforts have attracted attention and alarm, particularly in Europe. Indeed, one of the motivations for Britain's November 2001 emergency ban on reproductive cloning was the threatened imminent arrival of Dr Antinori. He had vowed to take advantage of a suddenly open legal loophole to clone some infertile British men, whose wives would carry the babies to term.

Approximately one year later, a similar call for a cloning ban in Serbia followed Antinori's visit to Belgrade.

Rounding out this cast of characters is Panos Zavos, a fertility specialist based in the United States, and for some time an associate of Dr Antinori. They had a rather acrimonious falling out, however, and Dr Zavos embarked on a

4. On offer: a clone of your own

cloning effort of his own. All of these projects are said to be taking place in secret laboratories in unnamed countries, with the mysterious locations and apparent lack of regulatory oversight adding to the consternation and concern.

Although some of these would-be cloners might seem like characters out of science fiction—or farce—the reality of cloning technology should be kept in mind. Human reproductive cloning is not a matter of if; it's a matter of when.

Sheep, cows, pigs, goats, mice, rabbits, horses, rats, a cat, and a mule have all been cloned by nuclear transfer, although no monkeys—not yet. Instruments necessary for the manipulation of single cells, including egg cells, are widely available in IVF clinics. There are thousands of such clinics worldwide, some in countries with little or no regulation. While the efficiency is very low, the technique of nuclear transfer is not all that difficult to learn. Indeed, a teenage girl, working as a summer intern at an American biotechnology company, was able to clone a pig.

All this being said, however, it is clear that the vast majority of IVF experts would not involve themselves in an effort to clone a human being now or in the foreseeable future. Given current levels of scientific understanding about the cloning process itself and the precise causes of the many abnormalities afflicting cloned animals, any attempt at human cloning is patently immoral and would result in ostracism of the scientist by his or her peers. As one practitioner told me recently, having a go at human reproductive cloning 'would certainly not be a good career move'.

The very thought of human cloning elicits fears that extend far beyond questions of safety. Unfortunately, cloning has become a proxy for more generalized fears about genetic engineering and about science and scientists out of control. In choosing his title for *Attack of the Clones*, George Lucas has shrewdly exploited these real-world associations.

After all, except for those of us who have met Dolly the sheep or any of the far less famous cloned denizens of various barnyards, fields, or labs, does anyone know a clone? Yes, in fact, virtually all of us do. Identical twins are clones—genetically identical persons—albeit produced by chance, not design.

5. Cloning in the garden

They are created naturally in the aftermath of sexual reproduction when an embryo splits into two identical embryos. Our real-world experience tells us that twins are just like us. Sharing a full complement of genes does not diminish their individuality or their humanity. One is not merely a replica or shadow self of the other. Lucas could have called his film *Attack of the Later Born Identical Twins*, but that doesn't sound very scary, does it?

Despite its heavy load of pejorative metaphorical baggage, the word clone had a benign and rather innocuous beginning. Derived from the Greek 'klown', it means twig. Taking cuttings of plants is cloning.

Thus, Britain, the land of gardeners, is also the land of cloners—of everything from apple trees to geraniums. This sort of purposeful horticultural cloning has been done for thousands of years, as have much more sophisticated efforts to ensure that plants were high yielding and disease resistant, to name but two valued properties. Some plants, such as elm trees, even clone themselves without our interventions, as their roots reach out below the ground to begin a new and identical life.

The word 'clone' is both a verb and a noun. As a verb, it refers to the process of creating a clone. Two different processes can produce clones: embryo splitting, the rare but naturally occurring event that produces identical twins, and nuclear transfer, the technique used to create Dolly the sheep. As a noun, 'clone' can refer to either the original animal or the 'copy'. For the sake of clarity, I will use the term 'progenitor' for the former and 'clone' for the latter. Cloning is a form of asexual reproduction—the creation of new life from a single parent. For many forms of plant and animal life, it's the natural way they reproduce. But it's not the natural way for mammals, and herein begins our tale.

On 5 July 1996, a new, noisy, and morally contentious being entered the world and transformed our experience of it. Dolly, the unsheepish sheep, was born in Roslin, a small town outside Edinburgh. She was the first mammal ever cloned from a creature that had already lived, and in her case, had already died. Dolly was cloned by nuclear transfer. The nucleus of a single mammary cell from a six-year-old sheep was inserted into an egg cell which had been emptied of its own nucleus, but still contained a gel-like substance called cytoplasm. Factors in the cytoplasm are believed to be responsible for reprogramming the incoming nucleus. This reprogramming is the essence of cloning by nuclear transfer.

Not much time and attention has been devoted to the extraordinary insights produced by cloning science or, indeed, to the morality of animal cloning. Instead, the worries articulated and the questions posed were then and are now about human cloning. Could we use these techniques to clone humans? Should we? Would we? Without waiting for the answers, legislators rushed to enact bans. Commissions were established to study the matter. Representatives of newspapers, magazines, radio, and television made the pilgrimage to Roslin to beg an audience with Dolly, who obligingly posed for photos, as long as she was given her favourite treat as a bribe—fortified grass pellets. Dolly remained a celebrity until her death on 14 February 2003.

The pilgrims who came from all over the world to visit Dolly were not the first the village had ever seen. Before her birth, Roslin's chief and perhaps only claim to fame was the Rosslyn Chapel. Built in the fifteenth century, it became an

6. The Apprentice Pillar

important destination for those who sought to glean from the still inexplicable iconography of the place the secret of the meaning of life. There are some very odd aspects to the carvings, which run the gamut from biblical to Masonic to pagan. It is said that some of the plant carvings depict flora from the New World, even though they were made before the discovery of America by Columbus.

The centrepiece of the Rosslyn Chapel is the extraordinarily beautiful Apprentice Pillar, the design of which is strangely reminiscent of the double helix structure of DNA.

Legend has it that the lowly stonemason who created the pillar was murdered by the mason in charge. His work was simply too beautiful, too perfect. For his accomplishment, born of pride as well as talent, he had to die. According to legend, his carved likeness stares down at his masterpiece for all eternity. Some say that the creation of Dolly was a similar act of hubris. That the secret of cloning from an adult mammal—enabling as it does a being to be reconstructed in the image of another from just a single body cell—was a secret better left unknown and undiscovered.

Hans Spemann, a brilliant German embryologist, conceptualized the Roslin experiment almost sixty years before Dolly was born, but he could not envision a way to transfer a nucleus from a differentiated cell into an empty egg cell. He termed his idea 'fantastic', and it was. Producing a cloned human being was the last thing on his mind. His aim was nothing less than unravelling a fundamental secret of developmental biology. The mystery of how an extraordinarily complex human being can develop from a single fertilized egg has perplexed scientists and philosophers at least as far back in history as ancient Greece. Aristotle performed his own investigations using chick embryos.

There were two possible explanations: either we were fully formed humans right from the beginning, and development merely involved growing larger, or we developed from the simple to the complex. Over time, scientists came to believe the latter theory. But accepting the premiss that a fertilized egg or zygote contains a complete blueprint for the formation of a human being raised a new question: is this master blueprint irretrievably lost as our cells differentiate and specialize to perform their required functions—becoming, for example, heart cells that beat, muscle cells that contract, and pancreatic cells that produce insulin? Or does the entire plan survive in every body cell, with all unneeded instructions languishing in a dormant or inactivated state?

Dolly's existence provided the proof, since validated by the cloning of other species, that, in differentiated cells, the entire body plan is suppressed but survives. The story of the unfolding of this fantastic experiment spans more than sixty years. It is a tale worthy of a great novel, full of dreams and disappointments, some successes, many failures, and even a couple of scandals.

Twelve years before Dolly was born, embryologists, also called developmental biologists, gave up the quest; they had decided that the fantastic experiment could not succeed in mammals. Cloning fell on hard times; it left the somewhat rarefied milieu of the academic laboratory and got shunted off to the barnyard. In a sense, it left the realm of the purely theoretical and entered the realm of the practical. Until the team at Roslin took up its work, Spemann's dream was kept alive by a brilliant, rather sardonic Danish vet, working in Britain with sheep, and by a team in the United States bent on producing masses of valuable cloned cows.

Cloning science is clearly fascinating in its revelations about the nature of development, but its importance goes far beyond that. Its applications have life-saving potential. Sheep, cows, and goats that have been genetically modified and cloned are already producing proteins in their milk that could be used to treat human diseases. At present, there is an unbridgeable shortfall between the supply of human organs available for transplant and the demand for those organs. Perhaps, one day, pigs which have been genetically engineered to decrease or even eliminate the causes of immune rejection by the human body might provide us with something close to transplants on demand—their own organs, so-called xenotransplants.

Could we clone pets? A cat has been cloned but despite prodigious efforts, scientists have not yet been able to clone a dog. If dog cloning proves to be possible, would Spot the Second even have the same spots as Spot the First? The answer is probably not. At least they might appear in different places. There is no reason to think that humans cannot be cloned. Would it be morally wrong? Most people say yes, but is this view a product of moral reflection or merely an intuitive response coloured by metaphors and misunderstandings? Are there morally sound arguments to support a ban on human reproductive cloning?

At least a temporary ban is morally obligatory. The efficiency of the technique in animals is extremely low and there have been abnormalities—some horrific—in clones from every mammalian species. Many have died before birth or shortly thereafter. Some animal clones that have survived gestation and birth seem normal. Perhaps they are, but defects too subtle to be noticed in a farm animal would be much more apparent in a human clone. Problems might not appear until long after birth, for example at puberty when genes that are supposed to get turned on remain switched off because of faulty gene expression. There has been much speculation about cloning and premature ageing. Dolly had arthritis and we are not sure why. On the other hand, cows cloned from senescent (rather old) cells appear to be as youthful as any other newly born calves.

What accounts for the abnormalities that we see in cloned animals? All our genes are present in every cell, but only the ones necessary to the particular function of each cell should be expressed (turned on). Perhaps, in clones, genes do not turn on and off in the proper manner, in the developing embryo and, after birth, in the cells of the animal. Reprogramming of the incoming nucleus by the cytoplasm of the egg may be faulty or incomplete. The very technique of growing cells in culture may also play a part in abnormal gene expression.

The cloning of Dolly was but the first of three extraordinary advances in human biological sciences that closed-out the old millennium and rang in the new. The second advance, the sequence of the human genome, was unveiled just as the new century began. It may well be that we are now at the crest of the wave of genetic determinism, sometimes called genetic reductionism—a view of what makes us who we are that comes down squarely on the side of nature in the long-running nature–nurture debate. Breathless announcements abound, announcing the discovery of genes that are said to explain even the most complex aspects of human behaviour including shyness, extroversion, risk-taking, and, perhaps most intriguingly, a bent for circus-performing!

Whatever one makes of any specific claim, we are clearly at the beginning of a long journey that will take us far beyond simplistic genetic reductionism—to an understanding of the workings of our genes and how they relate to each other and to our environment in a complex, intricate web of influences. Powerful ideas of genes as destiny have led people who have kept up on their reading to equate genetic identity with personal identity. Even a cursory acquaintance with a pair of identical twins should tell us that a clone with the same genetic inheritance as his or her progenitor would not be the same person. We are far more than the sum of our genes, as I hope will be clear as this book unfolds. The dream that cloning could allow us to achieve a kind of serial immortality is only that—a dream.

The third major advance in biological sciences was the derivation of human embryonic stem (ES) cells. It had taken almost twenty years to do in humans what Martin Evans and his colleagues at Cambridge had been able to do in mice. ES cells, unlike our body cells, which have become specialized, have the capacity to become any cell type in the body. Thus they have the potential to provide us with our very own body repair kit.

Cloning and embryonic stem cells come together in the technology of therapeutic cloning or cell nuclear replacement. The concept is simple but the scientific challenges are prodigious. A human embryo would be cloned from one adult cell united with an egg cell that had its own nucleus removed; then embryonic stem cells could be derived. These could, in turn, give rise to new therapies. Cells and tissues could be tailored for the person from whom the embryo was cloned. Because factors in the egg are the key to the reprogramming at the heart of cloning by nuclear transfer, both the research and clinical phases could be hampered by the already severe shortage of human eggs. Alternative strategies such as isolating the factors in the eggs themselves, retrieving eggs from aborted foetuses and deriving from them human embryonic stem cells are being explored. Eggs from animals such as rabbits and cows are another option, at least for the research phase.

Meanwhile, however, laws and policies to allow therapeutic cloning and even research on embryonic stem cells, derived from embryos left over after successful *in vitro* fertilization, and destined never to be implanted in any womb, have encountered fierce opposition from those who see such research as intrinsically immoral. Particularly in the United States, the horror and revulsion engendered by Clonaid's claim to have produced cloned babies have made the future of therapeutic cloning even more uncertain.

While most of us assumed that human cloning was biologically possible, nobody really knew that it could be done until February 2004, when a paper was published in the journal *Science*. In a virtual recipe book for those inclined to try their hand at reproductive cloning, a team of scientists from South Korea told us that they had cloned thirty human embryos and derived embryonic stem cells from only one. The researchers, who had practiced their technique with eggs from cows, were able to obtain an enormous supply of human eggs to use in their research.

Questions soon arose about the source of those eggs. We learned that some of them had been procured from female researchers who worked with the authors of the paper. Because the production of multiple eggs requires the administration of powerful drugs to stimulate the ovaries—a procedure that involves discomfort and at least some risk—morally troubling questions were asked (and still not satisfactorily answered) about the quality of the egg donors' consent. Coercion, if only the implicit variety, certainly cannot be ruled out.

The team achieved its breakthrough by using only one cell type—cumulus cells, which surround eggs in the ovary. Because these cells are only found in females, the Koreans were unable to clone males. Further narrowing the potential for either reproductive or therapeutic cloning, their technique only worked if the same woman provided both the nuclear material (from the cumulus cell) and the egg.

In August 2004 a team from the Newcastle Centre for Life was granted permission to undertake therapeutic cloining research in the United Kingdom. The British scientists are obtaining the eggs they need from women undergoing treatment at the Newcastle Fertility Centre. These eggs would otherwise be discarded. Scientists estimate that several hundred such eggs will be available every year.

Except for a very controversial experiment performed in the United States by scientists at Advanced Cell Technology (ACT) that yielded only a six-cell embryo, the British work is the first known effort to clone human embryos in Europe and, indeed, in the west.

Both cloning and stem cell research were made possible by advances in human *in vitro* fertilization (IVF) that took place in the 1970s, and culminated in 1978 with the birth of the world's first test-tube baby, Louise Brown. IVF has brought great happiness into the lives of so many people, but it has also created a morally problematic new construct—an early embryo developing outside the mother's body. Differing conceptions about the moral status of the early embryo are at the heart of raging disputes, which will not be resolved any time soon—if indeed they ever are. In Britain, therapeutic cloning is set to go forward under a clear and rigorous regulatory scheme. Policy makers in other countries have been far less keen, but as science advances and the clinical benefits come more clearly into view, attitudes and policies are bound to change, and the moral debate is certain to continue.

Both cloning and IVF became subjects of public fascination and moral consternation well before Dolly the sheep was introduced to an astonished world, more than twenty-five years before, as a matter of fact. Dredged up from our collective literary heritage were two novels, both of which continue to give their name to human interventions in reproduction and genetics—*Frankenstein* and *Brave New World.* In turn, these works inspired films and books which have made their own powerful contributions to the vocabulary of images and metaphors that we bring with us when we think of the reality of human clones. Creating life by cloning and other assisted reproduction technologies still seems to many to usurp and mock the role of God, devalue and diminish our essential humanity, and produce opportunities for totalitarian mischief. To those holding such views, the biological future looks bleak, even terrifying, and their moral angst is almost palpable. Science fiction has coloured our attitudes to the science facts of cloning. These fictions and the anxieties they elicit are the subject of the next chapter.

7. Dr Frankenstein and his monster

1

Power without Responsibility? Creating Life in the Laboratory

And God created man in His image.

Genesis 1: 27

I collected the instruments of life around me that I might infuse a spark of being into the lifeless thing that lay at my feet. . . . [M]y candle was nearly burned out when, by the glimmer of the half-extinguished light, I saw the dull yellow eye of the creature open: it breathed hard, and a convulsive motion agitated its limbs.

Mary Shelley, *Frankenstein*

One egg, one embryo, one adult—normality. But a bokanovskified egg will bud, will proliferate, will divide. From eight to ninety-six buds, and every bud will grow into a perfectly formed embryo, and every embryo into a full-sized adult. Making ninety-six human beings grow where only one grew before. Progress . . . [We can create] standardised men and women in uniform batches . . . Ninety-six identical twins working ninety-six identical machines. . . . The principles of mass production at last applied to biology.

Aldous Huxley, *Brave New World*

To lay one's hands on human generation [as has been done with in vitro fertilization] is to take a major step toward making man himself simply another of the man-made things.

Leon Kass, 'Making Babies'

A society that allows cloning, whether it knows it or not, has tacitly assented to the conversion of procreation into manufacture and to the treatment of children as purely the projects of our will. Willy-nilly, it has acquiesced in the eugenic re-design of future generations. The humanitarian superhighway to a Brave New World lies open before this society.

Leon Kass, 'Why We Should Ban Human Cloning Now'

Fictional creatures

The name of Frankenstein, the eponymous hero of Mary Shelley's novel, is known to most people throughout the Western world, irrespective of whether they have read the book itself. Augmented by decades of cinematic adaptations, the legend of Victor Frankenstein and the monster he created by assembling parts of dead bodies has become one of the central icons of our collective imagination. Myths and legends about the tragic consequences of over-weaning pride have a history dating back to the Greeks. Shelley pays homage to that heritage by giving her book the subtitle, 'The Modern Prometheus', alluding to the tale of the unfortunate fellow who stole fire from the gods and paid dearly for his offence. Fire, of course, is one of the basic requirements of civilization, so at least Prometheus was not punished for a trivial offence. Neither, of course, was Frankenstein.

Shelley departed from all prior versions of tales chronicling the nasty consequences of hubris in one crucial respect: there is no supernatural dimension—no tricks played on the gods, no bargains with Mephistopheles. Victor Frankenstein uses science and science alone to create life—without sex, without women, and, most of all, without God. For him, the scientific enterprise is all-consuming; it takes him away from human interactions and leads him to loathsome charnel-houses and dissecting rooms in his search for body parts from corpses. His very name has come to symbolize the mad scientist of popular imagination, hidden away and up to no good. He embodies and evokes the sum of our fears about the biological sciences: fears about scientists with impermissible motives, who conduct immoral experiments, and reveal secrets best left undiscovered; fears also that the power of science will strip us of our autonomy and dignity and render us less than human. As soon as he has animated his creature with the spark of life, Frankenstein is filled with regret, horror, and disgust at the consequences of his scientific quest. He fails to take responsibility for his creation and that, among other things, makes him a truly unsympathetic character. In fact, it is the monster who elicits our sympathy, having become evil only because he is spurned by humans who recoil at his ugliness. His acts of revenge are the physical embodiment and retributive mechanism of the evils of forbidden knowledge. As such, he destroys Victor and everything and everyone Victor holds dear.

In his power, emotionality, and primal force, Frankenstein's nameless monster embodies the Romantic movement's rejection of eighteenth-century rationalism and its hope that science could lead mankind to human progress and ultimate perfection. Victor's activities in the laboratory clearly take him in the opposite direction, to destruction, ugliness, and then despair. Cynicism and suspicion about extravagant hopes for science persist. The fear of modern biology relates not only to its power, but also to its promise and its very focus.

The focus is within; it's all about us. The scientific enterprise has been accused of robbing life of its mystery by revealing too much, by explaining too much. Now that the genetic text of life is being read, there is great disquiet. How can we continue to view ourselves as special if we share so many genes with chimpanzees, with mice, and even with yeast? If the locus of who we are is not in the heart, where is it? As the workings of the brain are better understood, what cherished romantic explanations will be vitiated? We certainly want the fruits of scientific knowledge—cures for dreaded and currently incurable diseases. But many of us are far less keen to have our genetic fortunes told. Most of all, we do not want the awesome power of science to fall into the wrong hands. Even our own. IVF babies, cloned babies, and notions of designer babies all conjure up visions of our being in control of reproduction—of our being able to bypass life's lottery and even the genetic lottery—and that control is at best a mixed blessing.

When Mary Shelley wrote *Frankenstein* almost two hundred years ago mechanisms for the generation of life were not yet understood. Therefore, she had Frankenstein create his monster mechanistically—by assembly. The idea that a human or quasi-human body could be built out of artificial components had important philosophical roots in the eighteenth century, when La Metrie wrote of man as machine. There are, however, literary antecedents that are hundreds and even thousands of years old in myth and legends about automata—pseudo-humans and fellow creatures that can trick observers into believing they are really humans. One of the most famous examples is from Offenbach's opera, Tales of Hoffmann (based on stories by the German writer E. T. A. Hoffmann). Olimpia is a life-sized doll, but she is so beautiful and she seems so real. Hoffmann is besotted; 'I know she loves me,' he cries, surely the ultimate in love-induced delusion.

There are many benign examples of manufactured creatures in fairy tales and other children's stories. Pinocchio, a toy, wants to be a real boy; the Velveteen Rabbit wants to be a real rabbit. In the film *AI* a robot specially made to be the much-loved replacement for a dead child literally goes to the ends of the earth to be loved and thus to be real.

Stories and films about manufactured life that are written for adults lack the benign and hopeful face of children's stories. Automata are the ancestors of the androids, cyborgs, and robots of modern literature and film. They can be produced in multifarious ways—by the gods (the Delphic oracles), by man using magic (Golem), or by man using science (Frankenstein's monster). For example, the golem, a figure from medieval Jewish literature and storytelling, is artificially created and then animated by magic so that he will look and act rather like a human. He can be helpful as a protector of Jewish villages or dangerous and hard to control.

Ideas about making people or people-like creatures entered the realm of mass production in Aldous Huxley's *Brave New World,* published in 1932. It takes place in AF 632, with 'F' referring to Henry Ford, whose Model T automobile was the first to be manufactured on an assembly line. God is dead or at least irrelevant. Under the banner of 'community, identity, stability', freedom, individuality, and authenticity are literally bred out of the lower castes of society through the twin mechanisms of ectogenesis (gestation outside the womb) and environmental conditioning (using chemicals, oxygen deprivation, and other means). The link between the mechanisms is the Bokanovsky process, cloning by embryo splitting, in a fantasy version capable of producing not the twins that occur in nature but ninety-six identical babies. The goal of the conditioning is creation of people with an inescapable social destiny—a destiny that they accept without question or complaint. In the Social Predestination Room babies with limited potential are decanted as 'sewage workers of the future'. The lineage of recent cinematic and literary cloned armies or drones can be traced straight back to Huxley's nightmare vision of mass production of nameless, faceless creatures with stunted human potential—produced to order for the purposes of the state.

Many people assume that the themes in *Brave New World* sprang full blown out of Huxley's imagination, but that is not the case. The ideas were in the air during the 1920s and writers including J. B. S. Haldane, Bertrand Russell, Lord Birkenhead, and Huxley's brother Julian all wrote about the themes and concerns that found their way into *Brave New World*. In these pre-double helix days, eugenics could only be achieved by selective breeding or sterilization of the unfit. The horrible fulfilment of such ideas about eugenics occurred in Nazi Germany.

Haldane leaned towards the optimistic view of the power of biological science to transform our lives. Russell was much more pessimistic: he feared that science could be misused to enhance the power of some at the expense of others. Haldane and Lord Birkenhead both predicted that the separation of sexual love from reproduction would characterize the human reproduction of the future. And, as IVF becomes increasingly more prevalent, it is clear that they were partially correct. Ectogenesis remains impossible—but for how long?

The work of ART in the age of mechanical reproduction

With a nod to philosopher Walter Benjamin for my adapting the title of his seminal treatise on aesthetics of some forty years before, in the early 1970s, the work of ART (assisted reproduction technology) in the age of mechanical reproduction elicited enormous anxiety. However, unlike the birth of Dolly the sheep, the birth of the first IVF baby, Louise Brown, did not come as a shocking surprise. Throughout the preceding decade, Robert Edwards and Patrick Steptoe, the principal scientists working in the field, issued regular updates about their research. As the public looked on in amazement, first animals and then humans were fertilized outside the body, in the proverbial test tube (actually a dish). Fictional dreams had become fact. Life could be created in the laboratory. One of the most private and meaningful events in human life had its privacy invaded and its meaning complicated, especially its moral meaning. What was this embryo growing in a dish? Was it a person? What could ethically be done to it?

Anxieties about the separation of sexual love from reproduction—evoked by and embodied in the mere mention of *Brave New World*—were soon joined by anxieties about reproduction (more accurately, replication) even without eggs and sperm. Seven years before Louise Brown was born, James Watson, the co-discoverer of the structure of DNA, published an article in an American magazine, *Atlantic Monthly,* called 'Moving Toward the Clonal Man: Is This What We Want?' The work of Steptoe and Edwards had made human cloning imminent, said Watson (one of the few scientists whose name was recognizable to the broader public), and that was a very, very bad thing. This article marked the beginning of the first wave of cloning anxiety and clear parallels exist between the public attitudes then and those that greeted the birth of Dolly the sheep more than twenty years later.

Philosophers, theologians, doctors, and lawyers involved in the fledgling field of bioethics took the question of cloning to their collective bosom. Dr Will Gaylin, the co-founder of what is today called the Hastings Center, a bioethics think tank, wrote an article for the *New York Times* Sunday Magazine explicitly linking cloning with the theme of Frankenstein: 'Frankenstein Myth Becomes Reality'. There was a flurry of scholarly papers exploring rather arcane arguments. The public was alarmed. Scientists stepped forward and Watson stepped back; the public was told that cloning wasn't really that imminent after all. So bioethics moved on to other topics that were more pressing in the 1970s, those that focused on the end of life rather than the beginning: the nascent field of organ transplantation and the first right to die cases.

Assisted reproduction remained a subject of moral concern but the focus shifted slightly. Once the first babies were born, it was hard to remain upset about the technique of IVF itself. Photographs of adorable babies don't quite mix with scary headlines. Instead, moral concern and public curiosity tinged with alarm were centred on the myriad of possibilities that conception outside the body made possible—including the separation of genetic and gestational parenthood. Leon Kass, the chair of President Bush's bioethics council, wrote several articles arguing against IVF; it was, he believed, the first step on the slippery slope to a brave new world in which children would be manufactured commodities. While he has since changed his mind about the morality of IVF, Kass has made many of the same arguments about the moral perils of both human therapeutic and reproductive cloning with liberal use of the Brave New World analogy.

Does the world of assisted reproduction really bear any resemblance to Huxley's nightmare vision? Central to *Brave New World* is mechanized reproduction for the benefit of the state. Biological engineering and environmental conditioning combine to serve totalitarian ends. Babies are produced in artificial wombs and then 'decanted'. There are no parents, there is no love. Humanity, creativity, and freedom are destroyed. All traces of individual identity are ruthlessly eradicated in the service of efficiency and communal cohesion. And perhaps most chillingly of all, people are manufactured to fulfil a predetermined role. There is no freedom, no autonomy, no chance to build a life of one's own.

When the latest innovations in assisted reproduction are reported in the media, 'Brave New World' is often used as a shorthand expression to suggest that we might be going in a rather scary direction. But, clearly, modern infertility treatment is nothing like the factories described in Huxley's book. People who cannot reproduce naturally seek assistance in order to have children to love and care for. Despite the separation of the sexual act from reproduction, what is going on is essentially what has always gone on: parents begetting children.

With respect to cloning, however, 'Brave New World' functions as much more than a shorthand term for headline writers. The book's nightmare vision of powerless, mindless, mass-produced clonal zombies fits all too well with contemporary ideas of clones and cloning. In a time when beliefs of genetic determinism are in the ascendancy, a clone, with a genome chosen for him by someone else, may seem to be as hobbled, constricted, and dehumanized as the products of *Brave New World*'s Predestination Room. In fact, as I hope this book will make clear, human clones would be nothing of the sort.

8. The nose

Cinematic creatures

It was not long before the cloning consternation of the early 1970s became manifest in literature and film. Works of fiction and films about duplicates appeared, starting in 1972 with Ira Levin's novel, *The Stepford Wives,* followed a year later by Woody Allen's film, *Sleeper* (a hilarious vision of the future, including an attempt to clone a dictator from his only surviving body part, his nose).

In 1976, the novel which has been almost as important to our collective cloning consciousness as *Frankenstein* and *Brave New World* appeared: Ira Levin's *The Boys from Brazil.* Although the book sold well, it is the film, released two years later, that has gained iconic status. The horror inspired by the story line was reinforced a thousandfold by using figures drawn from the Nazi era which had ended a scant thirty years before. Joseph Mengele, the so-called 'angel of death' of the Auschwitz concentration camp, is the main character. The real Mengele was fascinated by twins and conducted bizzare and cruel experiments on them in a quest to understand the relative contributions of nature and nurture. The fictional Mengele continues along similar lines by cloning ninety-four Hitlers. I will have more to say about this film later in the book, but suffice it to say here that socio-political uses of cloning technology were then and remain today far more disturbing to people than the idea of cloning as a medical treatment for infertility. It's not the cloned child that strikes fear into our hearts. Rather it's the idea that human beings could be designed to serve another's ends, to do another's bidding, to be controlled like a puppet on a string. Or perhaps even worse, that a quasi-human life form could be created that would be uncontrollable in its vengeance and destruction—like Adolf Hitler and Frankenstein's monster before him. It is not for nothing that modern fears of biotechnology and science spinning out of control are captured by the term 'Frankenscience'.

Two films about creating duplicates who were not quite human were released in 1978: a remake of the 1950s horror classic *Invasion of the Body Snatchers* and the film version of *The Stepford Wives*. We will explore the significance of these films in our discussion of identity later in the book. Four years later, Ridley Scott's *Blade Runner* was released. A disturbing and mesmerizing film, it depicts so-called replicants who live in off-world colonies. They are automata, assembled yet organic, that have been created to be 'more human than human'. They have implanted memories and bogus biographies. Some of them don't even know whether they are replicants or the real thing. To keep them from wreaking havoc, they are made with a genetic flaw, giving them a life expectancy of only four years. Like Frankenstein's monster, the replicants of *Blade Runner* resort to violence precisely because they have developed human feelings. They seek justice from amoral and irresponsible creators who have abandoned them to their fate. The replicants want longer lives; they come to earth to find their designer and demand freedom from their sentence of death. The film is suffused with ambiguity, even about the status of the narrator. Is he a replicant too? Probably.

The strangest cloning story of the 1970s has to be David Rorvik's 'non-fiction' work, *In His Image: The Cloning of a Man*. Rorvik, a journalist who had written a great deal about assisted reproduction, wrote about a human cloning experiment that he alleged had really taken place. He was not just the narrator but also the intermediary between Max, a rich and childless man who wanted to clone himself, and Darwin, an all too willing and rather mad scientist. Although the book was presented as fact, scepticism was immediate, particularly among the scientific community. One scientist, Deryck Bromhall, was particularly outraged. Fearing that he would be seen as the real Darwin and objecting to having his work on rabbit clon-ing cited in the extensive collection of endnotes of Rorvik's book, Bromhall sued and the publisher settled. Interestingly, Bromhall was involved in another work that year; he was the scientific adviser for the film version of *The Boys from Brazil*. His contribution is obvious. While the book discusses cloning by nuclear transfer (termed mononuclear reproduction) in a theoretical way, the film actually takes us into a laboratory and shows us how cloning would work—with rabbits. It was and remains widely assumed that *In His Image* was a hoax, though Rorvik continues to maintain that it was not.

The figure of Darwin, fictional or not, is squarely in the mould of Victor Frankenstein, exercising power without responsibility. *Brave New World* is ultimately a cautionary tale about totalitarianism; the threat to freedom does not come from science but from the perversion of science for ideological and political ends. In contrast, *Frankenstein* and its progeny are a direct attack on scientists and science itself. And because Mary Shelley's iconic protagonist's dream was to create a human out of parts of dead bodies, there is a rather pervasive belief that the scientists who cloned Dolly were just polishing up their technique to get it ready for human cloning. Such a belief is utterly false, as will become clear in the next chapter.

9. Portrait of Dolly

2

Reversal of Fortune: The Science of Cloning

[Experimental work is] far from being either the glamorous or dangerously uncontrolled activity many people imagine. It is also a great deal more laborious and time-consuming. The ratio of results to efforts is frighteningly small. It usually takes hundreds or thousands of tedious hours of work to obtain a result that can be described in a few minutes.

Lewis Wolpert

In an earlier life, I worked for the Marie Curie Memorial Foundation, a cancer research charity. I came to believe that we could use differentiated cells as donors for nuclear transfer because, in certain tumours, there are many different cell types that most probably originated from a single cell type. This suggested to me that the differentiated fate of cells is not fixed.

Keith Campbell

I think most people, once they have recovered from the initial shock, will recognise that the insights gained so far from mammalian cloning experiments have contributed in a positive way to our intellectual universe, whatever new shocks await us.

Steen Willadsen

Dolly and me

People all over the world were stunned by the announcement that a sheep named Dolly had been cloned from an animal that had already lived. Even many scientists were amazed, both by the nature of the breakthrough and the relatively obscure animal research facility where it occurred. I had my own particular reasons to be surprised. For, just one month after Dolly was born, about seven months before her existence was announced, I had visited the Roslin Institute for a meeting with Dr Ian Wilmut, the leader of the team that had cloned her.

Looking back on my experiences that day—and after the announcement about Dolly, I looked back on them quite a lot—I remember a reserved, soft-spoken scientist who apologized profusely for forgetting to book me a room in a hotel. Ian simply said, with what in hindsight was clearly colossal understatement, that he had been distracted by something. Exactly what, he did not say. I had come to talk to him about Megan and Morag, sheep that had been cloned at the Institute just one year before Dolly. This was a development reported with great excitement in the British press but virtually ignored in the United States. Megan and Morag had been cloned from embryo cells that had already begun to differentiate in culture—unlike Dolly, who was cloned from an adult cell—in a method that would lend itself to the main goal of the Institute. The Roslin scientists were working towards the production of genetically modified animals by a sophisticated technique they called gene targeting. The subject of our talk on that day in August 1996 was the implications that the Megan and Morag experiment held for xenotransplantation—the use of animal organs (specifically those of pigs) for transplantation into humans. We will explore these subjects in some detail in Chapter 3.

Ian gave me my first cloning lesson. He drew diagrams and explained the cloning process. Considering the secret he was keeping, there was nothing especially memorable about our talk except what he told me as I was leaving. In a quiet and very earnest manner he said, 'Our work here will revolutionize biotechnology.' Given the obscurity of the place and the radical disjunction between the modesty of the speaker and the extravagance of his claim, I remember feeling distinctly dubious.

In February 1997, when the British Sunday newspaper the *Observer* broke the embargo on a paper by the Roslin team that was set to be published in *Nature,* and the whole world heard of Dolly the sheep, I realized that Ian's final words to me that day were absolutely true. The fact that a differentiated or adult cell—one that 'knew' how to do or be only one thing (in Dolly's case, a mammary epithelial cell with the 'job' of producing milk proteins)—could be taught to recapture all the potential it had as an embryo cell (to be totipotent) was an extraordinary revelation to scientists. And there were equally extraordinary social and moral implications.

A mammal, in this case a sheep, but almost surely a human, could be reconstructed from a single cell of its body. And it was not even a germ cell, a cell which gives rise to eggs and sperm, but a rather run of the mill somatic cell, of which we have trillions! Although one might argue that Megan and Morag represented the true scientific breakthrough—the reversal of the differentiation that had already begun in embryonic cells dividing in a culture medium—the earlier experiment did not confront us with the prospect of reconstructing an existing organism from just one of its cells. But Dolly did. She also raised profound social and psychological issues in relation to the potential for cloning human beings.

One month later, Ian and I faced each other across a table at an office in London. He told me what his team had done and how his life had changed. A shy, retiring person had become the focus of worldwide media attention, and the quiet little institute had been overrun and overwhelmed. It seemed that everyone from everywhere wanted to meet Dolly. Several weeks later I did just that.

The Roslin Institute was a benign, suburban-looking sort of place made up of a small complex of low-lying, nondescript modern buildings with long, rather antiseptic corridors. Dolly lived in a barn not far away. I wondered what she would be like. Weren't all sheep passive, submissive creatures, indistinguishable from each other, and so dull that counting them was a widely used remedy for insomnia?

Indeed, the other sheep I came across at Roslin lived up to that boring stereotype, but not Dolly. She was brash, loud, insistent, spoiled, and she was fat. I had been told that she ran towards her visitors, not away from them. Indeed, Dolly stood up on her hind legs to greet me, and then ate out of my hand, 'really hoovering it up', as one of her caretakers noted. Afterwards, she posed (really, she did!) for a photograph. Dolly loved to eat and always had a weight problem. Once when I visited her, she was on a diet, mournfully staring and loudly complaining behind a small sign on her pen that read, 'Hay only'. Clones of several species have been overweight. Whether her girth was related to cloning or to a more prosaic combination of factors—being fed too much by charmed visitors and given insufficient exercise—is difficult to say.

10. Dolly and her surrogate mother

In February 2003, Dolly died as a result of a lung infection that had spread among cloned and non-cloned sheep alike. Compared with the life she led, in the glare of publicity, her death was rather a quiet affair. Incurably ill, with a future of pain and discomfort before her, she was euthanized. But Dolly will never be forgotten; she was a science pioneer—the first mammal cloned from an adult or differentiated cell of an animal who had already lived.

The procedure that had brought her into being, while

breathtakingly simple in concept, has proved very difficult to achieve. In her case it took 277 attempts. In order to create Dolly, the Roslin team performed nuclear transfer. A tissue sample was taken from the mammary gland of a six-year-old Finn Dorset ewe, whose identity the world will never know, and cells were grown up in culture. A nucleus was then removed from an individual cell and transferred into an egg from which its own genetic material had been removed. The egg was not empty, however; it was filled with cytoplasm—a gel-like material that occupies the part of the egg that is outside the nucleus. In egg cells, cytoplasm plays a crucial yet still poorly understood role in the cloning process.

An electric current was then applied, to fuse the nucleus and the egg and trigger the beginning of embryonic development. The resulting embryo was transferred into a surrogate from another breed, a Scottish Blackface. This was to provide visual evidence that the resulting lamb was in fact a clone (being of the same breed as the donor of the nucleus, but of a different breed from the surrogate).

Dolly was born on 5 July 1996. She was virtually an exact genetic copy of the six-year-old sheep that provided the nucleus. The only genetic differences were a result of mitochondrial DNA. Mitochondria are tiny organelles in the cytoplasm of the egg which produce energy. They were once independent organisms. About two billion years ago they in effect moved into primitive cells, establishing a symbiotic relationship, and thereafter evolved with their hosts into complex creatures such as ourselves. The bargain struck was that mitochondria, which themselves contain a small amount of DNA, get a home and the cells in which they reside get an energy supply.

Mitochondria are interesting because of their history

and also because of what they can tell us about our own. Since we inherit all of our mitochondrial DNA from our mothers, we can trace our own evolutionary path back through the female line. Some diseases are caused by mutations in the mitochondrial DNA and these are passed down from mother to daughter.

Because mitochondria are found in the cytoplasm (of the egg) rather than in the nucleus, there are small genetic differences between a clone and its progenitor. Although traces of mitochondrial DNA from the mammary cell that gave rise to Dolly have found their way into the clone, most of her mitochondrial DNA came from the enucleated egg, rather than the inserted nucleus.

According to a report issued by the US National Academies of Sciences, differences between a clone and its progenitor may manifest themselves in areas of the body that have a high demand for energy, such as muscle, heart, eye, and brain, or in body systems using mitochondrial control over cell death to determine cell numbers. These differences might be more obvious in human clones than in farm animals.

The diagram of the nuclear transfer process provided in Figure 11 shows all of the steps required in cloning a cow rather than a sheep, but the technique used in the two species is the same.

The puzzle of development

Scientifically speaking, the context of cloning is developmental biology, which seeks to unravel one of the profoundest of

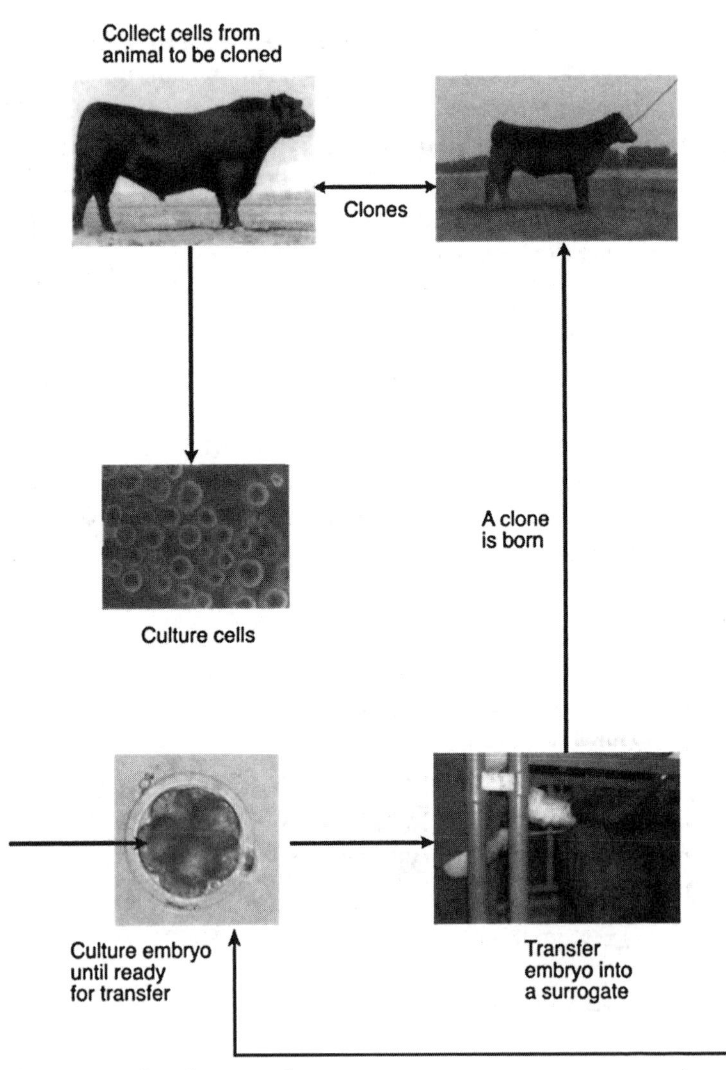

11. The process of nuclear transfer

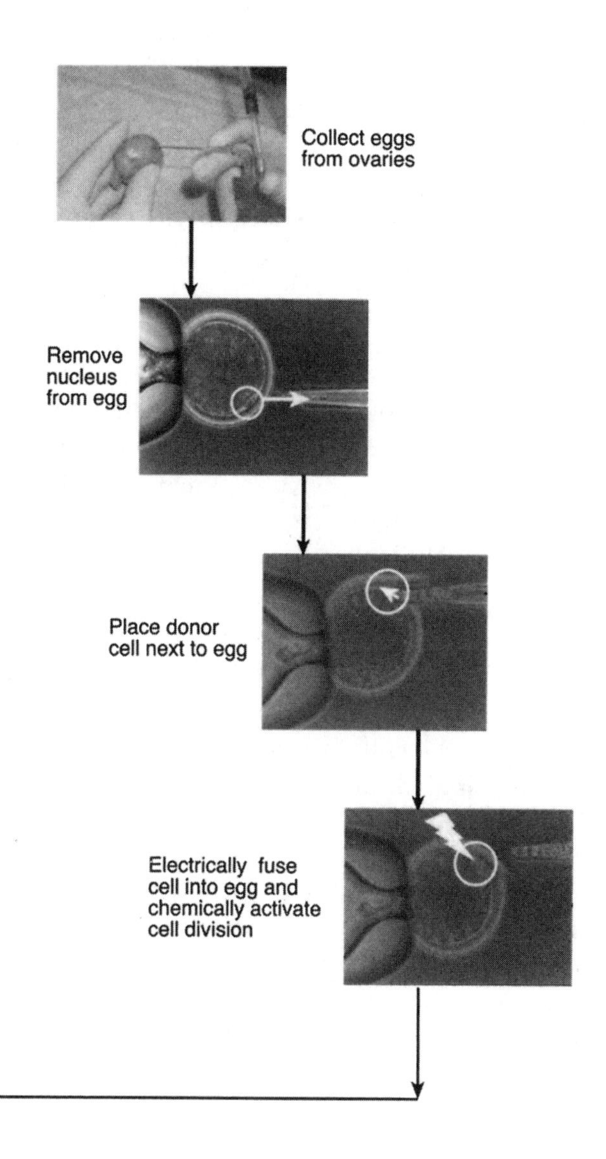
Collect eggs from ovaries
Remove nucleus from egg
Place donor cell next to egg
Electrically fuse cell into egg and chemically activate cell division

all mysteries—how a complex organism, such as a human being, develops from a single fertilized egg. To use a military metaphor: how do the cells, the foot soldiers of the body, organize themselves into vast armies? How are the commands given and received? What determines the strategy? How do the soldiers talk to each other? What signals do they use? One could employ a gentler metaphor—the orchestra. Then we would ask how all the different instruments learn to play just the right music at just the right time?

The mystery of development has both perplexed and fascinated scientists at least since the time of the ancient Greeks, and probably well before that. How do we get to be who we are? There were two prevailing explanations. The first, called preformation, held that all structures within us have existed from the beginning, from that very first cell. The process of development merely involved getting bigger. The visual symbol for this theory is the homunculus—a miniature embryo, or even adult, said to be hidden in the head of sperm cells.

The problem with this theory, aside from its inherent implausibility, is that if it were valid, all future generations would in turn be hidden within the sperm, like a set of nesting Russian dolls.

The alternative explanation held that we develop from the simple to the more complex, and from the general to the specific. Aristotle for one accepted this interpretation—and not just on faith. He performed his own experiments, breaking open chick embryos at various stages, in order to see whether new structures emerged with time. They did. The debate was not settled in ancient Greece; it went on for centuries. But those who believed that the plan for an entire chick or frog or human was there right from the beginning, in the

12. The homunculus

very first cell, had another question to answer: what happens to this body plan as cells specialize? Does it survive or are parts of the plan lost?

The essence of what Dolly had to tell us is that the plan survives. An adult or specialized cell, in her case one from the mammary gland, can be reprogrammed to start development—to have the same potential as that very first cell. But we are getting ahead of ourselves. Let's go back about one hundred years and meet the first of the giant figures of embryology who posited a theory about the fate of the body plan in specialized cells, a theory that several generations of scientists tried to and eventually did refute. His name is August Weismann, and he was a professor at the University of Freiburg.

Weismann is best known for two theories—one that has stood the test of time and one that has not. Indeed, his first theory—that we don't inherit acquired characteristics (such as physique developed by body-building, knowledge acquired through reading, or a skill learned, like tennis or chess)—marked an enormous advance in the understanding of development. He realized that during mammalian development, the reproductive cells (germ cells) are separated from ordinary body cells (somatic cells). What happens with respect to the latter is unable to be passed down to future generations through the former.

But it is his second theory about the fate of the original body plan that concerns us here. He believed that, as cells differentiate, parts of the body plan are lost, with the cells retaining only those genetic instructions required for them to fulfil their necessary and quite specific function. He thought that if one split a two-cell embryo, each cell would grow into one half of an embryo—one into the right half and the other into the left half.

There are all too few eureka moments in science. In the main, it is built on careful, painstaking, and often laborious work, and on the insights of others who have gone before. Newton's famous statement is certainly true: scientists stand on the shoulders of their predecessors. While there are those whose novel insights take the field forward in major advances, the mode of advance almost always involves asking questions, developing hypotheses to answer them, and designing experiments to test the theory.

The more exciting and revolutionary the hypothesis, the more motivation there is for scientists to test and even disprove it. Of course, scientists test their own explanations as well as those of others. But regardless of whose hypothesis is being tested, the mode of advancement is a dialectical one.

The story of Weismann's hypothesis illustrates this process very well. The obvious way to test his theory was to separate a two-cell embryo in order to see whether each cell was totipotent (capable of directing the formation of a whole organism). Amphibians were used because mammalian embryos develop inside the body of the female. There was simply no way, at least until *in vitro* fertilization (IVF) techniques were developed in the early 1970s, to observe and manipulate a mammalian embryo outside the body.

The fantastic experiment: Spemann and beyond

The first crucial experiments that challenged Weismann were performed in the late nineteenth century by Hans Driesch using sea urchin embryos. He was able to show that, after vigorous shaking, each cell of a two-cell embryo remained totipotent. This refutation of Weismann's hypothesis was then confirmed in other species by scientists including Hans Spemann, who later became the Director of the Kaiser Wilhelm Institute of Biology in Berlin and the only embryologist to win a Nobel Prize before 1986. In the last years of the nineteenth century Spemann was a young man, ill with tuberculosis, and in need of something to occupy his time. He read Weismann's book, *The Germ Plasm: A Theory of Heredity*, and it altered the course of his life. He decided then and there to become an embryologist. It was Spemann, at the very end of the nineteenth century, who continued the process of definitively refuting Weismann's theory, using a hair from his baby son—an ingenious albeit low-tech tool—to split a two-cell salamander embryo. Each cell developed into a whole embryo, not a half. Thus each cell had remained totipotent, at least at that early stage.

But in science, answers provoke new questions. The next question immediately suggested itself: did the cells of later-stage embryos retain the body plan? And then there was the most tantalizing question of all: when cells become fully specialized adult cells, do they lose forever their ability to direct more than their own narrow destinies? Spemann found a way to answer the former question by conducting the first, albeit very primitive, nuclear transfer experiment. Again, he used a baby hair but this time he managed to separate a nucleus of an older embryo cell from the cytoplasm that surrounded it and squeeze it into 'younger' cytoplasm. The cytoplasm was able to rejuvenate the nucleus and Spemann produced an identical twin of the original embryo.

Spemann was an extraordinarily gifted scientist. He won his Nobel Prize in 1935, and in 1938 he conceptualized the experiment that could definitively answer the question about the fate of the body plan in differentiated cells. In what he himself called a 'fantastic experiment', the nucleus would be removed from an adult cell and inserted into an enucleated egg. It was at once brilliant and audacious. But it was also impossible. Spemann had no means to insert the nucleus into the egg. It was to take another fourteen years before the cloning saga could continue through the experiments of Robert Briggs and Thomas King, working at the Institute for Cancer Research in Philadelphia. Briggs had a hypothesis that would explain differentiation and was keen to test it. He believed that the body plan was not lost and, instead, parts of it were simply turned off, with the unnecessary genes shut down.

Briggs and King worked with the embryos of spotted frogs called *Rana pipiens*. They designed and made their own tools and equipment that enabled them to suck the nuclei out of the frogs' egg cells and insert new nuclei from embryo cells. They produced the first clones by nuclear transfer—twenty-seven tadpoles. Their cloning of amphibians by nuclear transfer was not in any sense a trial run for a diabolical plot to clone humans. As with Spemann's work, Briggs and King saw nuclear transfer as a means of testing a hypothesis, answering a question and elucidating the process of differentiation. But scientific enquiry can be rather like peeling an onion; the more questions that are answered, the more new questions and challenges appear.

In the 1960s, Professor John Gurdon, working at Oxford University, took up the cloning effort where Briggs and King had left off, using another species of frog, the *Xenopus*—a mainstay, even today, for the study of developmental biology. Gurdon was able to get differentiated (intestinal epithelial) cells from tadpoles to develop into sexually mature frogs, but only with an extremely low efficiency. Most of the nuclei that he transplanted into enucleated eggs only gave rise to tadpoles. Neither he nor anyone else has been able to get differentiated cells from adult frogs to develop into sexually mature frogs.

Three cloned mice

This is where the situation remained until 1979, when a superstar of developmental biology, Karl Illmensee from the University of Geneva, announced that he had cloned mice by nuclear transfer from early embryo cells—the first time anyone had ever been able to clone a mammal. Illmensee, working with his colleague Peter Hoppe, published his results in 1981. He was by all accounts a brilliant and versatile scientist, and his lectures inspired at least two men, who were to become principal figures in mammalian cloning—Steen Willadsen and Keith Campbell.

The news that a mammal could be cloned—in particular one as crucial to the field of developmental biology as the mouse—caused great excitement among scientists. But there was a problem. No one was able to reproduce Illmensee's results, and it was alleged that he steadfastly refused to demonstrate his technique, even to those who worked in his lab. Rumours began to spread and several of his papers came under scrutiny. For there is a maxim in science that one result is a mere anecdote; good science should be repeatable. This issue came up again with Dolly because, even though many labs tried to clone a sheep from an adult cell for quite a long time, no one succeeded. Until the announcement from Hawaii in 1998 that mice had indeed been cloned repeatedly, there remained a great deal of scepticism as to whether a mammal really could be cloned from an adult.

Two scientists, Davor Solter, a developmental biologist who remains a giant in this field, and his student at the time, Frank McGrath, working at the Wistar Institute in Philadelphia, were the most relentless in trying to repeat Illmensee's results. Solter saw cloning, as did Spemann and Briggs, as a way to unravel the secrets of differentiation. He had come up with a strategy of his own for cloning mice, but after Illmensee's work was published, he abandoned his own method in favour of the technique that, by Illmensee's account, had proven successful.

But try as they might, Solter and McGrath could not get Illmensee's technique to work. After an exhaustive series of experiments and still no luck, they submitted their findings for publication. Although most scientific experiments end in failure—or perhaps because of that—journals generally do not publish papers relating negative results. But Illmensee's claim was so significant and controversial, and so many doubts had been raised about it, that in 1984 two prestigious journals, *Cell* and *Science,* published articles by Solter and McGrath. The last line of the *Science* paper appeared to sound the death knell for mainstream developmental biological research on mammalian cloning. The scientists wrote that their research suggested that 'the cloning of mammals by simple nuclear transfer is biologically impossible'.

It is still not clear whether Illmensee actually did clone mice. Keith Campbell, a key member of the team that created Dolly, thinks that he might have done it, but was unable to do it again. Certainly, Keith says, mice have been cloned since, so there is no reason in principle why Illmensee could not have succeeded. The impact of this scandal on the lives of Illmensee and Hoppe was profound. Far beyond this, however, was the effect of those memorable words in *Science* on the entire cloning enterprise. The most prestigious and respected developmental biologists simply gave up on the idea of testing Weismann's hypothesis and pursuing Spemann's 'fantastic experiment'. They moved on to other areas. Further enquiry was discouraged by simple neglect. All grant support for cloning research just dried up. It came to be seen as an endeavour unworthy of the great minds, and was literally relegated to the barnyard—the only place where there was motivation for getting it to work. Thus, cloning went from being a tool to advance knowledge to being a tool to accomplish practical goals, such as the reproduction of valuable animals. In the parlance of American baseball, cloning moved from the major leagues to the minors. It was in the domain of animal science—in the work of Steen Willadsen and Randall Prather—that the pessimism of developmental biologists was shown to have been misplaced. It was indeed possible to clone a mammal.

No one has done more to advance cloning science than Steen Willadsen, one of the most colourful of the cloning scientists. These days, he works mainly in the sphere of human *in vitro* fertilization. Born in Copenhagen, he grew up on a farm in Jutland and decided to become a vet. After working for a while, he became bored, and his restless curiosity led him back to university, where he received a Ph.D. in reproductive physiology. He then went to the British Agricultural Research Council Unit of Reproductive Physiology and Biochemistry in Cambridge, where he worked as a research fellow, funded by the Milk Board. His initial assignment was to develop a method of freezing sheep and cattle embryos, a project begun by his predecessor in the position—none other than Ian Wilmut!

Having successfully completed this project, Steen was free to choose his area of research; and he selected the micromanipulation of embryos. In the late 1970s, he devised procedures for developing identical twins and quads by separation and aggregation of cells from early embryos. In continuation of this work, he developed a complete procedure for cloning sheep embryos by nuclear transplantation, the mechanism that gave us Dolly.

Steen did extraordinary things with embryo splitting. For example, he split two-, four-, or eight-cell sheep, cow, goat, and horse embryos, grew the twin embryos in temporary sheep hosts and then implanted them in surrogate mothers of their own species. By freezing one of the twin embryos in a pair, he was able to produce identical twins of different ages. He also created chimeras—animals made by mixing together cells from embryos of the same species, and even from different species. He made sheep-goats.

Steen's next step was to investigate cloning by nuclear transfer. He transferred nuclei from eight-cell sheep embryos into enucleated eggs. In 1984, his first two cloned lambs were born dead but the next one survived. If Illmensee did not clone the first mammal by nuclear transfer, there is no doubt who did—Steen Willadsen. His sheep paper was published in March 1986. The year before, he left Cambridge for a ranch in Texas owned by the Granada Corporation. It was there that he adapted his embryo cloning procedure to cattle embryos.

When it came to cloning cows, Willadsen had rivals at the University of Wisconsin. They had commercial motives firmly in mind. W. R. Grace, a large American corporation, was interested in exploring cloning by embryo splitting. The goal was to exponentially increase the number of prized cows by cloning from eight-cell embryos. In effect an entire herd of cows could be produced from one embryo. The publication of

Davor Solter's paper in *Science*—with its bleak implications about the prospects of mammalian cloning—was bad news for the Wisconsin laboratory, which was headed by Neal First.

Randall Prather, at the time a young Ph.D. student, took over the project to clone a cow. The cow that he cloned in 1986 from an early embryo cell was born in 1987. There remains rather vigorous disagreement about who it was who cloned the first cow by nuclear transfer. Willadsen maintains that he did so, a full year before the team in Wisconsin. But it was the Prather team that published first. Prather's more recent and very exciting work involves knocking out genes in pigs cloned by nuclear transfer. With his work, the dream of using genetically modified pigs as a source of transplantable organs took a giant step forward. In Chapter 3, we will explore the way in which this work holds out the potential to transform the lives of everyone reading this book.

Even the proof that cloning of mammals from embryo cells could be done was not sufficient to revive interest in the developmental biology community. It was in the realm of commerce that cloning stayed and it was with commercial motivations in mind that the next big step in cloning science was taken. Megan and Morag were cloned at the Roslin Institute by Ian Wilmut, Keith Campbell, and their colleagues, the same team that, just one year later, created Dolly. How and why scientists at the Institute got into cloning will be described in the next chapter. Suffice it to say here that Keith and Ian performed two sets of experiments about one year apart. As a result of the first set, Megan and Morag were born in the summer of 1995. These sheep had been cloned from cells taken from nine-day-old embryos; the cells had been cultured and had begun to differentiate while still in the Petri

13. **Megan and Morag**

dish. The experiment proved that differentiation could be reversed and, because it did, the scientists at Roslin consider this first experiment to be the real leap forward from the point of view of science.

As we have seen, results in science, even positive results, tend to raise more questions. Why not, they thought, try to clone from cells that had differentiated even further—foetal cells and even adult cells? The second set of experiments performed at the Roslin Institute, the set that created Dolly, involved all three cell sources.

It was the creation of Dolly that captured worldwide media attention and unequivocally fulfilled Spemann's dream. If Weismann's hypothesis had been correct, differentiation would result in a progressive and irretrievable loss of potential. There would have been no way back to totipotency.

Dolly showed us that there was a way back from the restricted fate of adult somatic cells to the power and potential of the embryo. Briggs had been correct in his belief that, in differentiated cells, the entire body plan remained. The genes that were not needed to carry out the necessary and specific functions were merely inactivated or turned off. And even though differentiated cells only produce other cells that are just like them—muscle cells produce muscle cells, bone produces bone, skin produces skin, etc.—there is a way back. The cellular equivalent of a specialist can become a generalist, indeed, as with the cell that gave rise to Dolly, a universalist.

While some in the media did grab hold of the story of the Megan and Morag experiment and relate it to human cloning, many ignored it—especially in the United States. After all, these were embryo cells and, as one commentator said, why bother cloning if you don't know the clone will turn into Mozart?

Resetting the programme

The mechanism required to take a somatic cell back in time and restore the availability of the entire body plan is called reprogramming. Factors in the cytoplasm of the egg that are still not completely understood teach the nucleus of the donor cell to be totipotent. Inactivated genes become active once more. The nucleus can then direct the development of an early embryo. What remains unclear is whether the reprogramming-induced changes that render a mature cell totipotent can also turn back the clock of ageing. The answer is a resounding maybe!

Evolution did not equip the cytoplasm to do this massive overhaul of gene expression. When a sperm meets an egg in the process of fertilization, the cytoplasm of the egg does not have a big job to do; it has been designed to receive the incoming sperm. In contrast, when the cytoplasm has to reprogramme an adult nucleus in cloning, it is not only doing something that Mother Nature never intended; it is doing it at breakneck speed. Thus, it is believed, there is potential for the reprogramming to be incomplete or faulty in some way. The difficulties arise not just because of insufficient time, but because of the mind-numbing scale of the effort that is required.

The genetic programme that has to be reset is not restricted to events that happen over a short span of time—say early embryonic development. The programme operates throughout the life of the organism. For example, some genes that are important when a person reaches puberty are only turned on then. And when their job is done, they have to be turned off. If this intricate web of instructions is compromised, flaws in gene expression occur. Such flaws are termed epigenetic, to distinguish them from actual alterations or mutations in the genes themselves.

A somewhat useful analogy would be the electric lights in your home. When you are in the kitchen, you need the kitchen lights to be on. When you leave to go into another room, you switch them off. The crucial difference between the reality of epigenetics and my metaphor is that there are no adverse consequences, except perhaps a high electricity bill, if you leave lights on in a room after you have left it. But this analogy falls far short of conveying the complexity and coordination of what is required. The genetic equivalent of many lights must be turned on and off and at exactly the right time—rather like a dazzling light show perhaps. Cancer could be one consequence of leaving on certain genetic lights when they were meant to be off.

Although some animal clones appear to be healthy, most are seriously abnormal. Few survive to birth and, of those, more than 50 per cent die before puberty. The best hypothesis for the cause of the abnormalities that have been seen in at least some clones from every mammalian species derived from adult cells is faulty gene expression. A developing embryo seems to be extraordinarily capable of coping with imperfect gene expression—what scientists call noise. While gene expression does not have to be perfect in the embryo, nuclear transfer seems to produce rather too much of this noise, and epigenetic changes result. But it's not only faulty or incomplete reprogramming of the nucleus that may be responsible. Growing cells in culture can also cause epigenetic changes.

One specific kind of epigenetic phenomenon is called imprinting. Mammals such as sheep—and humans for that matter—inherit two complete copies of all the chromosomes in the genome. A subset of these genes, probably a very small subset, behave differently (are switched on or off) depending on whether they are inherited from the maternal or paternal line. One worry over cloning is that the reprogramming of the nucleus may upset the imprints and cause trouble. Perhaps not having parents from both sexes makes normal development impossible. Or perhaps because a clone comes from a body cell that itself has both a maternal and a paternal set of chromosomes—Dolly's anonymous progenitor had an equally anonymous mother and father—the appropriate imprint should still be there. We simply don't know.

14. **Dolly and Bonnie**

We do know, however, that Dolly developed arthritis, a condition that may or may not be related to cloning. She was also overweight. But, other than that, she appeared to be fine until she contracted the untreatable lung infection that led to her death. It is significant that she was sexually mature—a key factor because, as you will remember, no one has yet been able to use a differentiated frog cell to create a sexually mature adult frog. Dolly gave birth to six lambs, all produced not through cloning but the old-fashioned way and all epigenetically normal. I first saw her in her new maternal role in June 1998, just two months after she had given birth to her first lamb—named Bonnie because she was so beautiful.

Dolly had always been extremely friendly, but also rather highly strung. Not any more. Standing—quite proudly I thought—with her baby lamb, she was calm and quiet. I was the first person from outside the Institute to see her since her confinement, and the time away from staring strangers and television lights had given her a chance for a quiet life. She stood still while I petted her, and when I bent down to play with Bonnie, Dolly licked my face.

At long last—no longer the lone clone

For a considerable period after we learned of Dolly's existence, she was referred to and truly was 'the lone clone'. Because the scientific lesson that her very existence purported to prove was so important, there were doubters—lots of them. Dolly was called an anecdote, a pejorative term indicating that she represented a non-repeatable, and hence a non-verified, scientific finding. To make matters worse—or more doubtful—the data on the genetics of the source cells that gave rise to Dolly was far less complete than many would have wished.

Although the scientists at the Roslin Institute did not try to repeat this experiment—they had other scientific fish to fry, as I will explain in the next chapter—many scientists in other labs did try to clone a sheep from an adult cell and none

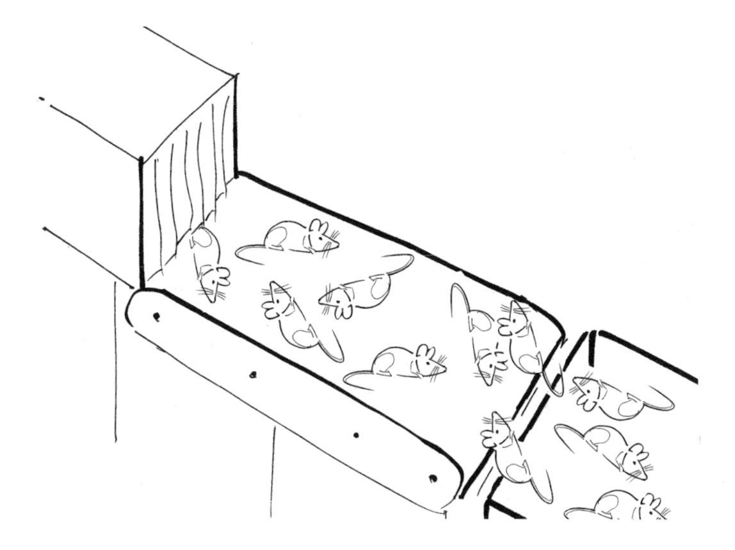

15. Cloned mice

succeeded. Then in July 1998, a paper was published in *Nature* that announced the successful cloning of mice by nuclear transfer. As with Dolly, adult cells were the donors. More than twenty of these mice developed into fertile adults. They were all female and the head of the family line was named Cumulina, in honour of the cumulus cells which surround each developing egg in the ovary. It was from this population of cells that the nuclei were taken.

The Hawaiian team was headed by a legendary figure in reproductive biology, Ryuzo Yanagimachi. The first author of the paper, Teruhiko Wakayama, has been described by other scientists as having magic hands. The technique he used was not the same one that produced Dolly. Instead of fusion of the nucleus with the enucleated egg, the Hawaiian team used nuclear microinjection. When Dolly was cloned, an electric charge both induced the fusion of the nucleus with the enucleated egg and initiated development of the resulting embryo. When the mice were cloned, chemicals were used to induce embryonic development.

The motivation of the scientists who cloned Dolly was, indirectly at least, commercial. In contrast, however, the mouse cloning work was done simply to explore theoretical questions surrounding cloning. Since this experiment was announced, male mice have been cloned, also by the Hawaii team, with the same technique but a different cell source: the tip of an adult mouse's tail. Subsequently, mice have been cloned from embryonic stem cells. I will tell you a great deal about the human equivalent of these cells, and their extraordinary potential for curing devastating diseases, in Chapter 4.

News of the first cloned mice has been followed by reports of cloned cows, goats, pigs, rabbits, horses, rats, a cat, and a mule. Efforts continue to clone other animals, including monkeys and dogs. Animal cloning can have profound implications for human health. What cloned animals offer us now and in the future is the subject of the next chapter.

16. Cloned pigs

3

Animal Farm: Cloning Applications

The main interest at Roslin was to find a method for the production of offspring by nuclear transfer from cells that could be maintained in culture and used as a route to precise genetic modification in animals for transmission through the germ line.

Keith Campbell

All animals are equal, but some animals are more equal than others.

George Orwell, *Animal Farm*

'Dr. Bryce. I'd like the transplant very much,' I said. 'It's quite simple really. I want to live.'

Malorie Blackman, *Pig Heart Boy*

Beware the Trojan Pig! When taken into your body it may release viruses hidden inside its organs just like the Greeks in the belly of the horse.

Robin Weiss

Banking on technology

Before Dolly, most of us thought about animal cloning—if we thought about it at all—in terms of *Jurassic Park*. As you may recall, the film tells the story of a rather unhinged scientist who brought dinosaurs back from oblivion to be star attractions at a theme park. He cloned them, using DNA extracted from fossilized mosquitoes (which had bitten dinosaurs) preserved in amber, an absurd scenario because fossils could not

17. Jurassic walk

contain sufficient intact DNA to allow cloning; there would just be too many gaps. With no dinosaur eggs lying about, he used eggs from amphibians and cloned only females, so that the creatures could not go forth and multiply. But something went horribly wrong and they did.

While we can't bring back the extinct creatures that once walked the earth, perhaps, with the help of genetic technology, we will be able to clone some of today's valued and valuable animals, such as much-loved pets and even species in danger of extinction. At this point in the development of cloning technology, a key element is DNA banking. In technique, it's rather like cryonics but in miniature. Instead of preserving the entire animal or person in liquid nitrogen (or, for the thrifty or body-conscious among us, just the head), scientists preserve only DNA that has been extracted from tissue. Of course cloning is also fundamentally different from cryonics; the former would produce a younger genetically identical twin, while the latter is meant to preserve the original animal or person in the hope that the medicine of the future could cure whatever illness or injury caused their death.

Teaching an old dog new tricks

In 1997, John Sperling, a Californian millionaire, had an idea: he would fund a team of scientists and ask them to clone Missy, his beloved dog. So began the Missyplicity project (www.missyplicity.com).

The scientists he funded were at Texas A & M University, a major agricultural school in Texas. As director of the project, Sperling selected Mark Westhusin, a well-respected sci-

18. The model pet

entist with a background in cattle cloning and an interest in dog reproductive physiology. At the time of writing, no attempts to clone Missy have succeeded. Mark did clone a cat—called Copy Cat. He was also able to clone a bull from an adult cell taken from Chance, a 22-year-old animal; the clone was named—what else!—Second Chance. In the summer of 2002, Missy died and, several months later, Sperling moved the Missyplicity project from Texas to an undisclosed location in California.

While the Missyplicity project has left Texas, Mark remains involved through a business venture set up by Sperling in 1997 to fund the animal cloning work. Genetics Savings and Clone is headquartered in an old bank building, perhaps even a now defunct Savings and Loan. The company offers DNA banking services to the public. There is standard service for healthy animals and emergency service for pets who are terminally ill or who have just died. Tissue samples taken by the client's own vet are sent to the facility via Federal Express. There is a yearly storage charge that will be deductible from the hefty cost once cloning services are ready for the market. And there is a backup facility, to ensure that the samples won't be lost in case of fire or other disaster.

If dog and cat cloning work well, produce healthy animals, and become affordable (the current price of $250,000 includes research and development costs), there could be a significantly large demand. It's no surprise then that several other firms, including Advanced Cell Technology (ACT), are offering similar services. Although people who expect a replacement pet to assuage their grief will certainly be disappointed, they will also see that clones are not alien monsters. Applying cloning to dogs and cats will give it a warm and furry face and make the prospect of human cloning far less frightening—perhaps even hastening its eventual acceptance.

The company's website (www.savingsandclone.com) is clear about what cloning a pet can accomplish; the clone will not be the same animal, it says. And then there are these words, 'if you truly believe that your animal's genetic endowment is so special that a newborn animal with the same endowment would also be exceptional—even if it did not know you or share any bond with you—then gene banking your late pet may be the right thing for you to do.'

Pets and people are valued for special qualities that result from a complex and subtle interaction of genes, environment, and experience. At least some of those so desperate to resurrect dead pets or dead people or replicate existing ones hope for some continuation of consciousness. They hope that cloning would produce a younger version of the same person or dog or cat. This dream is just as impossible as the 'repet' scenario in the Arnold Schwarzenegger film *The Sixth Day*, in which Arnold and his daughter's dog are cloned as fully formed copies, ready to assume the identity of the originals.

The grass menagerie

Dolly was cloned from a progenitor who had already lived, but she was not cloned to replicate a special sheep; indeed, her progenitor has faded into obscurity, leaving behind only a chunk of frozen mammary tissue. The advantage of cloning from an identifiable animal's somatic (body) cell is predictability, genetically speaking. Thus, Dolly set the stage for cloning animals from prototypes with desired characteristics—such as better milk production for cows, and endurance and speed for racehorses. Age differences apart, the progenitor and the clone or clones would be virtually interchangeable. Unlike pets, farm animal clones are not created to be loved but only to be used or to be sold.

When scientists at the University of Wisconsin, working with funding from the W. R. Grace Corporation, cloned their cow by nuclear transfer in 1986, they were engaged in a very practical enterprise: production of multiple versions of valuable animals. Because they used early embryo cells—and not cells taken from an adult animal—the scientists did not know in advance the precise characteristics of the resulting animals. But, because the embryos came from prize livestock, they had a very good idea of what to expect.

In 1994, W. R. Grace sold its American Breeders Service unit, ABS Global, to a private equity fund. Three years later, ABS Global became Infigen, a small biotechnology company. In October 2000, scientists there cloned Mandy, a valuable Holstein cow; the clone was sold at auction for more than $80,000. When I visited Infigen, I saw many cloned cows grazing in the fields. Unlike the flocks of cloned sheep and litters of cloned pigs I had seen—all of whom were indis-

19. Cow clones

tinguishable from each other and from non-cloned animals—the cows had very specific markings. From a distance, they looked like multiple images of the same animal. It was quite an eerie sight, so I asked for a photograph.

After many attempts in several labs, horses have been cloned. Clones of champion racehorses would be extraordinarily valuable animals. Already, their sperm can fetch a great deal of money—often much more than what they earned from racing. Cloning would be an even better bet. What would it be like to watch the Grand National or the Kentucky Derby if the entire field consisted of clones of the horse that won the race say five years before? Their DNA would be virtually identical—except for about 0.5 per cent located outside the nucleus—in the mitochondria. In terms of horse racing this tiny difference might actually matter; mitochondria are the mini power stations of the cell so perhaps they could account for small differences in physical strength and endurance. As the animals would be raised in different environments (beginning in the womb), we would not only see a horse race, but also a fascinating nature–nurture experiment. Such a race could also be the acid test for the relative skills of the uncloned jockeys. One thing is certain: predicting the outcome would give the bookmakers fits!

Attempts are being made in several countries to clone species about to become extinct, for example, the Spanish bucardo mountain goat. Some of these experiments require using an egg and a surrogate mother of another species and this can cause problems. The American company ACT cloned a rare Indian ox called a gaur, using both a cow egg and surrogate. They named him Noah. He was born alive, but succumbed to a common infection after only two days. Perhaps one day pandas, which breed poorly by conventional means, could be cloned, or even rare tigers.

Rewinding the clock?

The very existence of Dolly and other mammals cloned from somatic (body) cells with narrow destinies and functions proved that, developmentally speaking, cloning renders the nucleus of donor cells young again. The potential of youth is restored. But what about youth itself? Is a clone born old or is the ageing clock turned back along with the developmental clock? What experiment could scientists design in order to find out? Before we take up those questions, a bit of background seems in order. Somatic cells increase in number and replenish themselves through cell division. Unlike germ cells—eggs and sperm and their precursors—they are programmed to die after they divide a specified number of times. The exact number appears to differ depending on the cell type.

Within each of our cells, the physical structure that carries the genetic material or DNA is called a chromosome. At the end of each chromosome is a bit of DNA that does not code for (give rise to) proteins. These ends are called telomeres. They are rather like shoelace tips or the leader at the beginning of a roll of film. Every time our somatic cells divide, the telomeres shorten. As long as enough of the telomere tip remains to keep the coding portion of the chromosome from fraying, cell division can continue. Eventually, however, the telomeres become too short to protect vital portions of the chromosome from damage. At that point, the cell stops dividing and dies.

Would the telomere length of a young clone's cells be the same as that of its progenitor? Scientists at the biotechnology company ACT designed an experiment to find out. They cloned twenty-four cows from a senescent (old) cell and then measured the telomeres of the calves. They published their results in the 23 November 2001 issue of *Science*. An old cell with shortened telomeres had produced cloned calves with longer telomeres. The clones were born with cells no older than those of calves born the old-fashioned way. The ageing clock—at least at the cellular level—had been reversed.

In another series of experiments involving cloned mice, their telomeres were also shown to be longer than those of their progenitor cell. An American team led by Teruhiko Wakayama cloned six successive generations of mice. They made clones of clones in order to discover what happened to their telomeres. In a paper published in *Nature* (about a year before the ACT paper was published), they reported that telomere length increased slightly with each generation.

The results of mouse and cow experiments came as a great relief to the animal cloners. Clearly, if cloning from an adult progenitor produces a clone that is old before his time, it would be a lot less attractive as a means of duplicating prize animals. And, of course, human cloning would have one more moral strike against it.

But what about sheep? What about Dolly? Her progenitor was six years old when she was cloned. There has been a great deal of speculation—especially after it was announced that she had rather premature arthritis—that Dolly might be ageing rapidly. Did her lifespan start from six, rather than from zero? When Dolly's telomeres were measured, they appeared to be 40 per cent shorter than expected for a sheep of her age. This very different finding may result from species differences or, as Keith Campbell thinks is possible, it may stem from the experiment itself. The scientists at Roslin only measured the telomere length in Dolly's white blood cells. Perhaps, Keith says, she had some kind of infection.

Because telomere length serves as a marker for cellular ageing, scientists have long wondered whether they are also a marker for the lifespan of the entire organism. Would an animal or person with lengthened telomeres have a greater life expectancy? It is certainly an intriguing possibility but there is no scientific evidence—at the moment—that would support it.

Although clones do not appear to be born with old cells, many of them are abnormal in other respects. Animals of every species cloned thus far have exhibited severe abnormalities. There have been miscarriages, still births, and early deaths. Some clones do appear to be normal but the procedure also remains extremely inefficient. As we have seen, the most likely cause of the abnormalities are epigenetic flaws caused by incomplete or inaccurate reprogramming of the inserted nucleus by the factors in the cytoplasm of the recipient egg. The culturing process may also play a role. Cloning has produced lambs that could not catch their breath—unable to propel their blood through enormous blood vessels that were twenty times larger than normal. Autopsies have revealed shrunken kidneys and undifferentiated liver cells unable to do their proper job. These are but three examples from the catalogue of defects. Large offspring syndrome, which afflicts cloned sheep but not other cloned species, is also seen in sheep produced by IVF, but the abnormality in size is much greater among the clones. The incidence of abnormalities and their severity argue very strongly against any attempts to clone human beings; we neither understand the causes of the problems nor have the capacity to prevent them.

Cloning animals for simple replication opens up a variety of scenarios—some fascinating and fanciful, others rather prosaic. A different sort of animal cloning will prove much more important in the future. Cloning in combination with genetic modification with one or more genes from another species (transgenesis) holds out extraordinary promise with respect to human healthcare. There are two broad categories to consider: cows, sheep, goats, and rabbits, already being engineered to produce valuable proteins in their milk or blood; and pigs modified to become suitable source of organs for human transplantation. Both categories can be produced by first- or second-generation transgenic techniques. While the second generation is upon us, the first still has useful applications.

Pharming

The first transgenic animal, a mouse, was created in 1982 by transferring a gene from one animal to the embryo of another. Ralph Brinster of the University of Pennsylvania and Richard Palmiter of the University of Washington were recognized for their work by the French Academy of Sciences, which awarded them its highest honour, the Charles Leopold Mayer prize. During the 1980s, scientists in Cambridge, led by Martin Evans, developed an ingenious method for introducing genetic changes in mice; they used mouse embryonic stem cells. We will spend quite a lot of time discussing stem cells in the next chapter—in reference to the extraordinary potential of their human counterparts for developing treatments for hitherto incurable human diseases.

The Roslin scientists were interested in transgenics long before they had ever contemplated cloning. They produced what I will call first-generation transgenic animals by a simple if not always effective technique: injection of the foreign genes directly into the embryo of the animal. I will use the term second-generation transgenics to refer to very precise and targeted genetic modifications made possible by the cloning process itself. Before we explore this aspect of the work at Roslin and Dolly's place in the scheme of things, I want to introduce you to three transgenic animals that might seem more at home in the fevered imaginings of science fiction writers than they would be grazing at a farm near you.

Could goats prevent malaria—one of the scourges of humankind? Perhaps, one day, a single herd of some very special goats could provide sufficient vaccine to eliminate malaria from the entire continent of Africa. Scientists have engineered goats to produce an experimental malaria vaccine in their milk. They did this by introducing a transgene from a deadly malaria parasite into a goat embryo. The transgene was designed to be switched on by the cells that line the mammary gland of the goat. The World Health Organization estimates that malaria infects three hundred to five hundred million people a year and kills one million. First-generation transgenic methods are being used in this case.

Could goats produce spider web silk in industrial quantities for uses as diverse as medical sutures and bulletproof vests? Again, the answer is yes and the means is transgenics.

Scientists at a small Canadian biotechnology company have injected spider genes into goat embryos and produced goats that secret a protein called fibroin in their milk. Once the fat is removed, the protein can be spun into fibres that are, pound for pound, five times stronger than steel and more elastic than rubber. First-generation methods are being used here as well.

And, finally, could cows function as pharmaceutical factories and even neutralize biological weapons? A team of American and Japanese scientists have cloned transgenic cows by adding human genes for immunoglobulin—a substance in our blood that is central to the proper functioning of our immune system. Most transgenic animals have been engineered to express only one specific foreign protein. In contrast, because these cows express what might be called the raw material of our immune response, scientists should be able to vaccinate them against a wide range of deadly pathogens and elicit an immune response to order. The cows would then produce a plentiful supply of antibodies that could be used to treat a wide range of diseases such as smallpox, anthrax, and botulism. An infected person—or a large population of persons—could be spared the risk of side effects inherent in vaccines and also the time lag before they take effect. Immunity conferred by immunoglobulin-derived antibodies would be immediate. These human antibodies would be especially valuable with respect to botulism since there is no vaccine that can prevent it. The potential for this tech-nology extends beyond bioweapons defence—to providing an inexpensive and plentiful range of treatments for conditions as diverse as transplant rejection, leukaemia, and auto-immune diseases.

Back to Roslin

In 1987, the Roslin Institute formed a partnership with PPL Therapeutics, a biotechnology company just across the road. Their goal was to produce transgenic animals that would secrete pharmaceutical proteins in their milk. They were especially interested in a substance called alpha-1-antitrypsin (AAT), a protein that could be very useful in the treatment of cystic fibrosis. The scientists knew their goal; the crucial question was how to get there.

Dolly and Megan and Morag were not the first celebrity sheep at the Roslin Institute. Tracy, rather a local celebrity in comparison, was born there almost twenty years ago. She was famous for her ability to secrete large amounts of AAT in her milk. While she was able to do the job that needed to be done, she was the product of inefficient and rather undependable first-generation transgenics. Injection of genes into an animal embryo is inefficient and it can also cause problems. Genes integrate into their new environment randomly and, if a vital gene is disrupted, the embryo could die.

A new and better method of gene modification had to be found and, for this, Ian Wilmut, Keith Campbell, and their partners at PPL looked to nuclear transfer as a vehicle. The experiment that produced Megan and Morag proved that nuclear transfer could be done with cells grown in culture—cells that had already begun to differentiate.

The last step to be taken called for the addition of a human transgene while the cells were in culture, and then nuclear transfer to produce the animal. The process is basically the same as that used to clone Dolly and Megan and Morag, with the crucial difference that genetic changes are

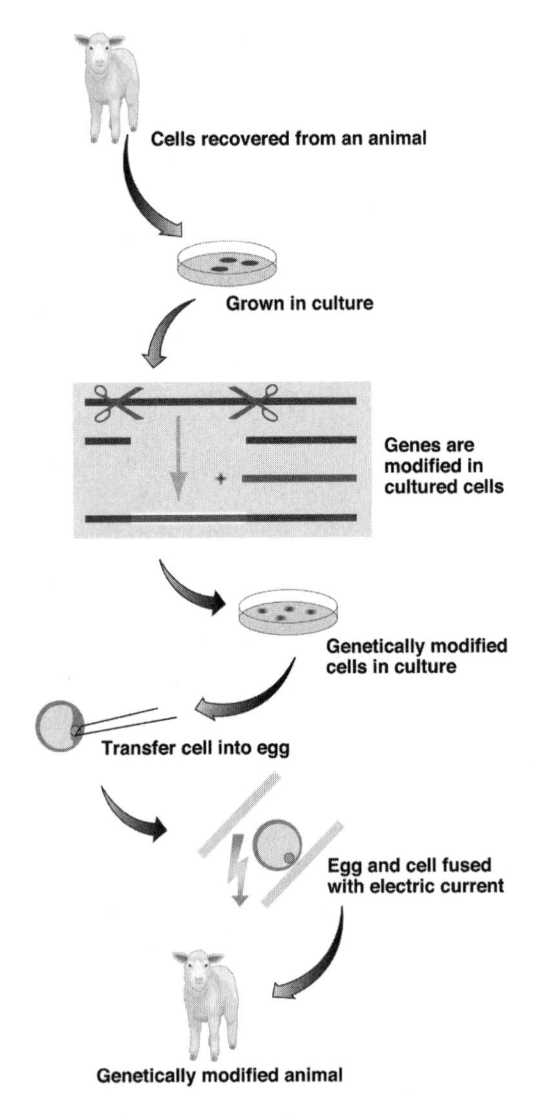

20. How to make a transgenic animal

made in the donor cell population before nuclear transfer takes place.

For the Roslin scientists, cloning was not a form of mass production of animals. Nor was it—as it was for embryologist Hans Spemann, Briggs and King, and the others—a means to learn whether the original body plan survived in somatic (body) cells. And it certainly wasn't a trial run for cloning humans. Nuclear transfer for Ian Wilmut and Keith Campbell was simply a means to a very practical and commercially useful end.

Dolly was not part of the overall strategy of the Roslin scientists for two reasons: she was not transgenic and she was cloned from an adult cell. Ian and Keith believed that foetal fibroblast cells would better lend themselves to genetic modification while adult cells, such as the mammary cell that gave rise to Dolly, would be useful primarily for duplication of valuable animals.

It is an enormous irony that Dolly—the sheep who veritably shook the world—was not integral to the scientific or commercial strategy of the Roslin scientists; she was a mere detour on the road to transgenesis by way of cloning technology. Commercially speaking, their goal was a very different sheep: Polly, a lamb born in July 1997, exactly one year after Dolly.

Polly was cloned from foetal fibroblast cells that had been genetically modified in culture. A true designer sheep, she had a human gene in every one of her cells. While designer sheep could well provide us with plentiful and inexpensive pharmaceuticals, designer pigs could do them one better—providing organs suitable for transplantation into humans.

21. **Polly and her sisters**

The pig in the cupboard

Not long after Dolly's existence became known, the tabloid press went wild with speculation about the clone in the cupboard scenario—people cloning themselves to create brain-dead doubles, whom they then keep in reserve for spare parts. Needless to say, growing replacement organs inside human bodies is nobody's idea of a solution for the transplant crisis.

Growing them inside non-human bodies, however, is another matter. The clone in the cupboard has no future in medical research, but the pig in the cupboard very well might. In our second category of genetically modified animals, pigs for transplantation, scientists followed exactly the same journey as their counterparts at Roslin. Before we accompany them along their extraordinary path of discovery, some background seems a good idea.

Human organs for transplantation are becoming more and more scarce. Surgical techniques are improving; indications for transplantation keep broadening; and fewer people are dying in ways most suited to harvesting their organs. Motorcycle helmet and seat belt laws as well as better trauma care have markedly decreased the most suitable source of human organs: young, healthy people who die suddenly as a result of a blow to the head.

Xenotransplantation is the transplanting of cells, tissues, and organs from one species into another. Somewhat idiosyncratic attempts to graft animal parts in humans date back to the seventeenth century, when a dog bone was reported to have been used to repair the skull of a Russian aristocrat. In the 1920s, Serge Voronoff, a Russian scientist working in France, transplanted slices of monkey testicles into his very rich patients to achieve what one might call the Viagra effect. There is a long history of attempts to transplant animal organs into humans. Until the adoption of brain death criteria about thirty years ago, it was impossible to transplant a viable human organ without murdering the donor. If one waited until the heart ceased to beat on its own, it was simply too late; the quality of any organ obtained at that stage was far too poor for it to be useful.

With respect to xenotransplantation, three basic questions arise: will the animal organ work in the human body? Will it be rejected by our immune system? Will it cause infection in its human host? The principal danger posed by any transplant is rejection—the recognition by the body of a foreign invader and the activation of an immune response. The response to an animal organ is much stronger than it is to an organ from another human. While organs from non-human primates, such as monkeys, evoke a stronger reaction than a human organ, a transplant from a more distant species, such as a pig, elicits an even more vigorous response, termed hyper-acute rejection (HAR). Within minutes, the organ turns into a black, swollen, useless mass.

Although primate donors have the advantage of genetic similarity and therefore pose relatively less risk of rejection, they pose a higher risk of harbouring diseases that might be transmitted across species. For a variety of reasons, pigs are much more suitable, but, as we will see later, transplanting pig organs into humans would also involve viral risks. Their organs are the right size, they breed quickly, they have been domesticated and raised for food, so moral issues about killing them for human benefit are not as problematic as those raised by using primates. Primates, however, are still required for research purposes and it is important that they be treated well and killed in a humane manner. Pigs should, of course, also be treated humanely and their organs should not be removed serially.

In common with the sheep at the Roslin Institute, the journey towards genetic modification of pigs began with microinjection into embryos in order to add human genes. In 1992, David White, a British scientist working at Imutran, a

22. Spare parts

small biotechnology company in Cambridge, created the first transgenic pig by injecting human genetic material into pig embryos. The organs to be transplanted were thereby coated with human proteins and, in effect, able to trick the human immune system into mistaking them for human organs.

Imutran and its scientists had become targets of animal rights extremists. When I went to visit David and his pigs, he was in his office but the pigs were hidden away in a secret location in rural Cambridgeshire. Imutran itself had extremely tight security. There were guards and strict sign-in procedures. And the windows were mirrored and made of bulletproof glass.

David and his colleagues had been able to defeat HAR, but they still had a major problem to solve. The monkey recipients of modified pig organs required toxic doses of antirejection drugs to keep them from succumbing to the ravages of acute vascular rejection (AVR), which seems to be a continued albeit less severe manifestation of the same problem that causes HAR.

The best hypothesis for the cause of both types of rejection involves the workings of a pair of genes in the pig (one each from the maternal and paternal line) that code for an enzyme called alpha 1,3 galactosyltransferase. The enzyme adds the sugar alpha gal to the surface of pig cells. Alpha gal coats the surface of mammalian organs—except in humans and old-world monkeys, both of whom lost the enzyme in the course of evolution. Thus, the human body would reject a pig organ as soon as it detected the sugar coating. While adding human genes can mitigate the rejection, scientists have been hoping that deleting or knocking out the pig genes might solve the problem completely.

The next step

As was the case with the Roslin sheep, before scientists could do precise genetic modification of pigs, they had to master the task of cloning them by nuclear transfer. In 2000, scientists at Infigen, working in association with Imutran, published the first paper documenting the cloning of pigs. Imutran soon closed up shop and moved its operations to North America. There were many reasons for the move but the more favourable animal research climate was a prime consideration. In September 2000, Imutran was merged with part of US-based BioTransplant into what is currently known as Immerge Biotherapeutics. Scientists at BioTransplant had been working in close collaboration with Professor David Sachs, Director of the Transplant Biology Research Center at Massachusetts General Hospital. He had been breeding a herd of so-called miniature swine for more than twenty years. When I went to see him, I was rather shocked at the size of the pigs. Although they were about one quarter of the size of some of the Imutran pigs, they still weighed 250 pounds! Their size makes them more appropriate sources of organs for humans.

In order to genetically modify pig organs for transplantation into human recipients, the Imutran and Immerge scientists followed the same three steps—albeit for very different purposes—as their counterparts in Roslin: micro-injection of human genes; cloning; and, finally, precise genetic modification accomplished through nuclear transfer.

They injected human genes into pigs in order to make them more acceptable to the human immune system; this was the essence of the work by David White. Then they had to clone pigs by nuclear transfer. This was achieved by colleagues at Infigen. Their last step was truly a giant one: instead of using gene targeting to add human genes in culture, they used the technique to delete pig genes.

For this work, they collaborated with Randy Prather, the scientist who was the first to publish his work on cloning cows by nuclear transfer. In 2001, he was lead author of a paper in *Science* describing the creation of a single knockout pig. In December 2002, scientists at PPL published a paper in Science describing creation of double knockout pigs, with both copies of the gene that produces alpha gal having been deleted.

The viral risk

The recipients of genetically modified pig organs would, of course, be humans. Safety, then, is the threshold scientific and ethical issue, even if the rejection problem is solved. The great unknown is the degree of virological risk. In a 1997 paper in the journal *Nature Medicine*, Robin Weiss, then at the Institute of Cancer Research in London, demonstrated that pig endogenous retrovirus (PERV)—which had been lurking in the germ line of pigs for millions of years—could reappear in infectious form and infect human cells in culture. Because the virus is integrated into the genome, it cannot be bred out. But perhaps it could be knocked out through gene targeting techniques made possible by nuclear transfer. This may be difficult, however, because there are so many copies of PERV in the genome.

Because of the risk of viral transmission—not just to patients but to the public at large—we must proceed with caution and with great attention to the adequacy of the human recipient's informed consent. Patients, in dire medical straits, may be tempted to overlook the risks of being a recipient of a pig organ. They might be burdened psychologically by having a pig organ implanted. Because of concerns about pig viruses, they, and probably their family members, would be subjected to a lifetime of surveillance by public health authorities.

Many hurdles remain but, one day, pigs may provide spare parts for humans. But what if we could avoid the risks of infection and rejection by growing our very own body repair kit—using our cells as raw material with which to produce healthy cells and tissues and, perhaps one day, even organs? The promise of therapeutic cloning and stem cell research is the subject of our next chapter.

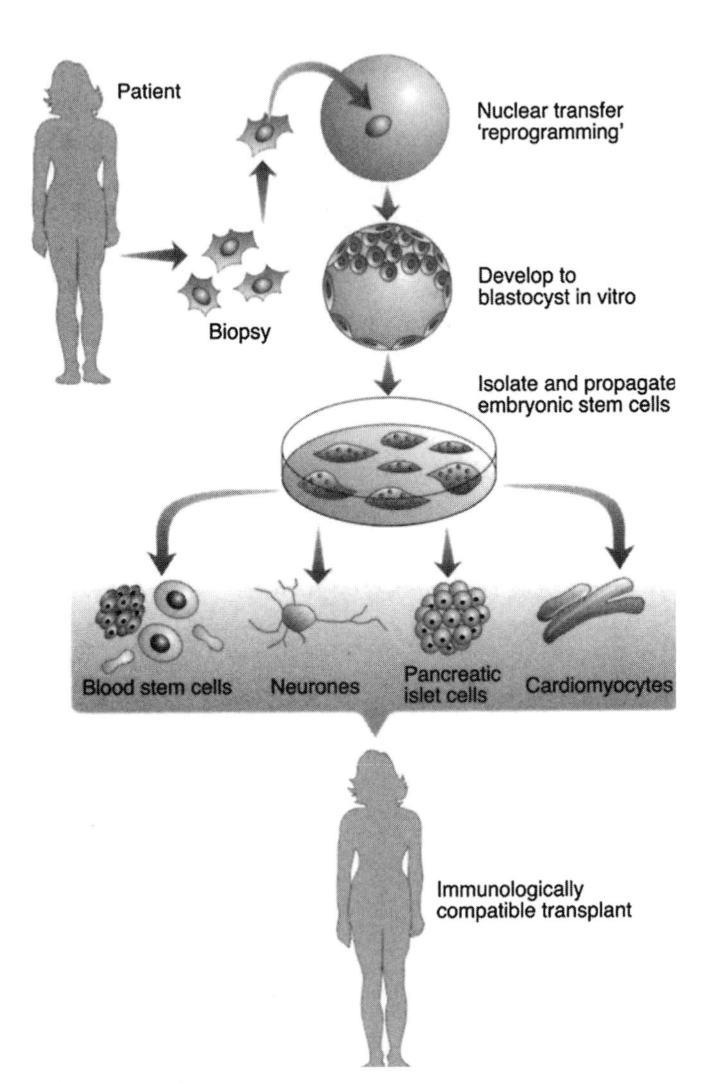

23. Therapeutic cloning

4

Building Your Own Body Repair Kit: Cloning for Cell Therapies

Some people are opposed in principle to all forms of embryo research on ethical grounds. But we must also recognise that when stem cell research has huge potential to improve the lives of those suffering from disease, there are also strong ethical arguments in its favour.

Tony Blair

Every honourable member was once an embryo . . . [L]ife begins at conception.

Edward Leigh, Member of Parliament for Gainsborough

What's in a name? that which we call a rose
By any other word would smell as sweet.

William Shakespeare, *Romeo and Juliet*

If some other way of describing the procedure could honestly have been used, I believe that many of the difficulties in which we find ourselves would go away. Even with the qualification of the word 'therapeutic', the word 'cloning' sends shivers of horror down the spines of the British public.

Mary Warnock, Member of the House of Lords

Tainted?

What is true for roses may not be true for some medical technologies. In the case of cloning, as Baroness Warnock has noted, the word itself is laden with such negative metaphorical baggage that it tends to conjure up horror and revulsion in many if not most people. Such a reaction is certainly evoked by talk of cloning by nuclear transfer for the purposes of human reproduction. There is also a medical application on the horizon. The nucleus of one of a patient's adult cells could one day be reprogrammed in order to derive cells for therapy. Unfortunately, this potentially life-saving technology has been tainted, perhaps even fatally so, just by the name it has been given—therapeutic cloning. Indeed, in the United Kingdom, another term for the technique is often used in an attempt to avoid these negative associations—cell nuclear replacement (CNR).

Let's see exactly how this would work in the future.

In the first stage, a biopsy is taken from a patient, from skin perhaps, and the cells grown up in culture. Then the nucleus from one of these adult cells is transferred into an egg which has had its own nucleus removed. Factors in the cytoplasm of the egg reprogramme the nucleus by erasing the regulators of gene expression characteristic of adult tissue. In a sense, these factors trick the nucleus into behaving as if it came from a one-cell embryo. The embryo is then allowed to develop for about five days, to the blastocyst stage, when it contains about 150 cells and is smaller in size than a grain of sand.

At this point, embryonic stem cells (ES cells) can be derived from what is called the inner cell mass. They can then be grown up in culture, multiplied, and rendered immortal—preserving their potential to become all other cell types.

Means must be found to coax these cells into differentiating along predictable and necessary pathways so that cell types required for the patient providing the biopsy are produced. Once these cells are created, they can be transferred back into the patient without rejection and, hopefully, provide a treatment or even a cure for his or her disease. I will explain what stem cells are, how they function, and how they could be the source of life-saving therapies later on in this chapter.

Conceptually therapeutic cloning is quite similar to a procedure with which many of us are already familiar—an autologous blood transfusion. Several years ago, I had elective surgery. As a precautionary measure, in case I required a blood transfusion, I donated a pint of blood for myself. If a transfusion had become necessary, my own blood would have been given back to me. If therapeutic cloning had been possible then and I could have benefited from its use, cells provided by me could have been transformed into other cells, all genetically identical to mine, and these transferred back to me in a medical procedure.

Therapeutic cloning is a two-stage process. In the first stage, a patient's differentiated cell—which has already realized its destiny and is only able to perform narrow and specific functions—is taught to direct the reconstruction of an entire organism, to become a one-cell embryo. Differentiation is reversed and is then resumed. This stage is identical to the initial step in reproductive cloning. The key to the second stage of therapeutic cloning—realizing the promise of embryonic stem cells—also relates to differentiation. It involves learning how to direct these master cells with an utterly open future down specific pathways, so that they can be transformed into the cells needed for specific therapies.

ES cells produced by therapeutic cloning hold out an enormous advantage over their other possible source—spare or surplus embryos left over after IVF treatment. ES cells derived from cloning embryos would not provoke the bane of transplantation medicine—immune rejection. In as much as the cloned embryo would be genetically identical to the patient whose differentiated cell gave rise to it, so would the stem cells. And so would cells and tissues produced from these stem cells. Thus, toxic drugs that prevent organ rejection would not have to be administered, saving the patient from side effects including increased susceptibility to both infection and cancer. It should be noted, however, that scientists are pursuing strategies other than CNR to get around the problem of rejection and that, to date, there is no evidence in humans that therapeutic cloning will actually work.

While the goal of therapeutic cloning is certainly a clinical one—getting treatments to patients who desperately need them—even at the research stage, therapeutic cloning could be an extremely valuable learning tool. Through research, scientists could learn to understand the reprogramming that is involved in nuclear transfer so that, one day perhaps, we could bypass the embryo stage entirely and, say, take a cheek cell from Mr Jones who has Parkinson's disease and turn it into neurones (brain cells) that will cure him. Or heart cells, called cardiomyocytes, to provide a patch for Mrs Smith's damaged heart. Thus, if scientists are able to unlock enough secrets of this process, therapeutic cloning could one day even put itself out of business.

Despite the similarities of therapeutic and reproductive cloning in their initial stages, in both their intention and in their product they could not be more different. I will come back to these differences—and their moral significance—at the end of the chapter. Meanwhile, let's try to shed cloning's metaphorical baggage and concentrate on the *raison d'être* of therapeutic cloning: the extraordinary promise of the embryonic stem cell.

Hope to end the suffering

No medical technological wonder has fired up expectations of scientists and the public the way stem cell research has. ES cells are derived from microscopic early embryos. These cells have the potential to become any cell type and thus to replace tissue that no longer functions due to disease or damage. If scientists find the right growth factors to enable them to be predictably and reliably coaxed along their developmental way, we could, one day, rebuild our failing organs, cell by cell, and perhaps, using tissue engineering, rebuild our bodies organ by organ.

Human ES cells lend themselves to metaphorical description. They conjure up images of genies granting wishes, chameleons being transformed, the elixir of life. All of these phrases are apt, and more besides. Right now, for so many people, these cells seem the very stuff that dreams are made of. Primal stuff. In the full flower of their youthful malleability, they have been derived from the early developing embryo and rendered immortal in a dish. There, with proper nurturing, they will continue to divide and grow while scientists try to unlock the secrets to how to make them do our bidding.

It is impossible not to be moved by the sight of 'superman', the American actor Christopher Reeve, imprisoned inside a body that no longer works. A riding accident left him paralysed below the neck. He has become a passionate and articulate advocate for stem cell research because he believes that stem cell-derived therapies will be used in tissue repair and, perhaps one day, allow him and others like him to walk again. Probably as a result of an intensive physical therapy regimen, he has made amazing progress in his struggle with paralysis. Alas, progress on the political front seems far more elusive. America is falling behind other countries. There is good reason to believe that the first human trials of stem cell-derived therapies will happen in Britain, and Reeve has said he will cross the Atlantic for treatment. He wants to get himself back, as do millions of sick and suffering people.

The key to his hopes and those of so many others is the quintessential spare part, the ES cell. The best guess at the moment, and it is only a guess, is that stem cell-derived therapies may reach the clinic in five to ten years—perhaps initially for blood disorders, cardiovascular disease, or Parkinson's disease and other neuro-degenerative conditions. Progress depends greatly on the resources, intellectual and financial, that are available for research. Also crucial is the regulatory environment because polices can nurture research (as they do in Britain) or they can stifle it (as they have done and seem destined to continue to do in the United States).

Stem cell science

Stem cells—whether they are derived from an early embryo or adult person—have two key properties. Under the right conditions, they have the ability to reproduce themselves for long periods of time (the life of the organism, in the case of adult stem cells). They can also give rise to the specialized cells that make up the tissues and organs of the body. This second characteristic sets them apart from most of our bodies' cells—those making up our skin, muscle, or liver, for example. If such cells are able to replace themselves at all, they can normally only produce other cells just like them.

Scientists believe that most stem cells are able to give rise to only a limited number of cell types. Adult stem cells are found in many of our organs but not all of them. Those that are unipotent can give rise to one type of specialized cell; those that are multipotent can give rise to all the specialized cell types of an organ. Adult stem cells are found in human bone marrow, blood, both the cornea and retina of the eye, in the brain, skeletal muscle, and the pulp of our teeth, among other locations. Their job is to replace and replenish cells with more specialized functions, such as muscle cell contraction or nerve cell signalling. Think of them as an organ's local repair facility. Under laboratory conditions, rare bone marrow cells and neural stem cells have formed other cell types and repopulated tissues of the body that are damaged or diseased.

In contrast, pluripotent stem cells are able to give rise to the three germ layers, mesoderm, endoderm, and ectoderm, from which all the cells of our bodies are derived. There are only two known sources of such cells—blastocysts (embryos about five to seven days old), and the primordial germ cells (the ancestors of sperm and eggs). More recently, a subset of mesenchymal stem cells (from bone marrow) and neural stem cells have been shown to have apparent pluripotent abilities. There are over two hundred different cell types in a human and ES cells may one day be able to produce all of them. In a sense, to order.

We know the potential of ES cells but we don't yet know how to make use of that potential. Human embryonic stem cell research is a very new field. It was only in 1998, almost twenty years after ES cells were isolated in mice by scientists in the UK, that two teams of US researchers, sought out and funded by visionary scientist Michael West, published papers revealing that they had succeeded in isolating human pluripotent stem cells. Dr James Thomson and his group at the University of Wisconsin, Madison, derived human ES cells from the inner cell mass of spare embryos (left over after IVF treatment) and grew them into five immortal cell lines. Dr John Gearhart and his team at Johns Hopkins University in Baltimore, Maryland, isolated embryo germ cells from primordial reproductive cells obtained from aborted six-to-nine-week-old foetuses.

Soon papers began appearing—and not only about ES cells. Scientists were discovering the surprising chameleon-like feats of adult stem cells. No one had believed that these cells were capable of versatility beyond their range of tasks in situ. As often happens in science, conventional wisdom is yielding to new insights. Research using adult stem cells continues and it is showing extraordinary promise—both in the context that would come naturally to these cells and in other more unexpected ways.

They can do what comes naturally and do it very well. For example, injection of neural stem cells into the spinal fluid of paralysed rodents has restored movement. They can also do what does not come naturally at all. Although adult stem cells have already become specialized and thus would be expected to have a rather narrow destiny, several recent studies have shown that, in some cases, these commitments can be overridden and that one type of adult stem cell can be reprogrammed into another type. This malleability seems to be particularly true with respect to nerve and blood stem cells.

A useful analogy, provided in an article published in *Science,* compares the feats of adult stem cells to a music student becoming a successful professional baseball player. Scientists are teaching these cells to alter their course, to reverse their fate. The task with respect to embryonic stem cells, however, is quite different. Scientists don't need to make them alter their natural behaviour. They just need to direct that behaviour in desired, therapeutically useful directions.

Adult stem cells may provide a perfectly adequate basis for the treatment of some diseases. However, when compared with ES cells, adult stem cells seem to have very real shortcomings, such as poor accessibility, those in the brain, for example, and limited supply. They have not been identified for all cell types. They also bear the consequences of ageing so we would expect them to have accumulated genetic damage (mutations).

Until recently, there had been no evidence put forward to indicate that any type of adult stem cell has the ability to differentiate into every type of cell in the body. A paper published in Nature in June 2002 by the University of Minnesota scientist Catherine Verfaillie described a very rare type of adult stem cell that she and her team had isolated from human, mouse, and rat bone marrow. Like ES cells, these cells were able to multiply indefinitely in culture and, injected into mouse embryos, the mouse cells were able to differentiate into most cell types. The full potential of Professor Verfaillie's discovery is yet to be explored but many scientists think it unlikely that her cells will prove to be as versatile as ES cells.

If scientists could transform a patient's adult stem cells into the cells needed to cure disease or damage, we could have our very own body repair kit without the need for therapeutic cloning. There is a lot at stake and we are still at the beginning of a long and very exciting journey. From the point of view of the science, research should proceed on all fronts.

Getting stem cells into the clinic

While public concern has thus far been limited to the moral pitfalls, there are scientific pitfalls as well. The stuff of life is very finicky stuff indeed. It can metamorphose in a chaotic sort of way—with little teeth, tufts of hair, and heart cells that beat all developing in the same dish. The scientific hurdles will be challenging. We not only have to learn how to teach embryonic stem cells to turn into what we do want. We have to make sure they don't turn into what we don't—to ensure that they don't produce a cure that is worse than the disease.

We want a metamorphosis with an endpoint: production of stable cells. What we don't want are new heart cells that suddenly veer off and become liver cells; or nerve cells becoming bone; or liver cells becoming nerves. Such switching of fates after transplantation would not only be weird; it would be deadly. Perhaps scientists will insert a suicide gene that would kill the cell if it started to metamorphose yet again, the molecular biological equivalent of the spy's trusty cyanide capsule.

Before stem cell-derived therapies can enter the clinic,

they must be shown to be safe from contamination by pathogens and also from infiltration by other sorts of cells. Early human embryonic stem cells were cultured using a feeder layer composed of mouse cells. Because mice have retroviruses which integrate into the DNA of cells, the discovery in 2002 that ES cells could be produced without mouse cells is crucial for clinical applications.

Embryonic stem cells themselves would not be transplanted. They are the raw material, not the finished product. They are much too similar to cancer cells, which have two main characteristics: they have reverted to a more primitive, less differentiated state than normal cells in the organ where they arose, and they grow in an uncontrolled way. As stem cell therapy nears the clinic, great care must be taken, and it will be.

One of the earliest applications may well be Parkinson's disease. Its physical and psychological toll at present is enormous. An estimated 500,000 Americans are afflicted with Parkinson's disease, with about 50,000 new cases diagnosed yearly. In the UK, there are approximately 120,000 sufferers. When patients' families are figured into the equation, the numbers affected in the two countries are about 4 million and 1 million respectively. All of these people are living with the consequences of this terrible disease on a day-to-day basis. Drug treatments can be effective, but only in the short term. And there are common adverse side effects, including hallucinations, involuntary movements, and the sudden and unexpected onset of sleep.

A surgical technique to inject foetal cells into the brains of Parkinson's sufferers was developed in the 1980s, but British patients have not been offered this treatment for about ten years. This is because four to ten foetuses are required per patient and both the required number of foetuses and the cost of the procedure were judged to be too high. The procedure has been offered in the United States, but it has been mired in controversy—first moral and then scientific.

Heart failure may be another promising target for cell therapies. The shortage of human hearts for transplant is intractable. This is the motivation for the xenotransplantation research that I described in the last chapter. Another way beckons—using embryonic stem cells to derive cardiomyocytes, the muscle cells of our hearts that keep it beating. Perhaps doctors would be able to intervene early, before the heart is on its metaphoric last legs, and make patches for failing hearts—repairing them and arresting their decline. This could, one day, lessen or even obviate the need for human hearts for transplant. There are so many people who are suffering, and alleviating that suffering is surely a moral imperative. But what about the embryo? Is it ethical to conduct embryo research?

A new moral construct: the embryo in the dish

Almost twenty years to the day before Gearhart and Thomson published their papers on stem cells, Edwards and Steptoe, working in Cambridge, gave the world its first test-tube baby, Louise Brown. They also gave us a new biological and moral construct: the early embryo developing in vitro (in a dish). While many if not most early embryos produced in vivo (in the body), without the need or benefit of IVF, are simply lost, this loss is hidden from view. It just happens without our even knowing. Early embryos developing outside were very different. They could be observed, manipulated, frozen for later use, and, perhaps most problematically, doctors could decide whether or not to implant them. They knew what they could do to these embryos, but what ought they to do?

Four years after the birth of baby Louise, the Warnock Committee was created in Britain to examine questions surrounding assisted reproduction and the embryo research that would be required to assure the safety and increase the efficiency of the technique. As Mary Warnock observed recently, while discussion of the moral status of the embryo is usually phrased as an enquiry into just when life begins, that is not the question at all. We should more properly be asking when the embryo becomes morally significant. Is the early embryo already a person, with the same rights as a newly born baby, or is it more akin to other cells and tissues? Is the moral calculus different with respect to embryos *in vitro* and foetuses that are developing in the womb?

A question of life

The Committee's report, *A Question of Life,* was published two years later. Six years after that, its conclusions were incorporated into the United Kingdom's Human Fertilization and Embryology Act of 1990—legislation that regulates embryo research and assisted reproduction with donated gametes. The Committee's principal conclusion was that, although the early embryo possessed a moral status rendering it deserving of respect, that status did not afford it absolute protection. An early embryo could not be considered a person; thus the moral duties we owe to persons simply do not apply. Research should be allowed but only until the embryo reaches the fourteenth day of development and the primitive streak appears. At this point, the body plan of the embryo begins to be laid down. One consequence of this is the induction of the nervous system.

Because IVF was and remains inefficient, the creation of surplus or spare embryos became a crucial part of the technique. They could be stored for five years, subsequently extended to ten, and if neither used nor donated for research, they had to be destroyed. Embryos could also be created solely for research purposes but this has been done very rarely; during the first twelve years that the Act was in force, just over one hundred research embryos were produced.

The Human Fertilization and Embryology Act allowed embryo research for purposes involving IVF, congenital disease, miscarriages, contraception, and diagnosis of chromosomal abnormalities. In 2001, eleven years after the Act became law, regulations approved by Parliament allowed embryo research for the additional purpose of developing therapies for serious diseases. The case was made in Parliament on behalf of those who were suffering grievously and could benefit from the research. Their moral claims to be heard and to be helped simply outweighed the case presented by the pro-life campaigners for the absolute sanctity of the very early embryo. ES cell research would proceed in Britain.

Cynical science?

The great majority of the British and American public do not believe in the absolute sanctity of the early embryo. But there is a passionate minority that holds the belief that life begins at conception and, therefore, all embryo research, beneficial or not, is morally wrong. This belief is shared in several European countries. Pro-life advocates on both sides of the Atlantic often attempt to make a scientific case rather than a moral one. Early on in the public debate—certainly long before the Verfaillie paper was published—they could be found praising the value of adult stem cells.

Tactically, this is quite a clever strategy. When one runs up against the stone wall of opposition in a morally fraught political cause or line of argument, there are only two choices. A moral argument can be put forward that would appeal to a limited audience: those who already agree or at least are disposed to be brought around. Or one can disguise deeply held yet not widely shared moral and religious beliefs in a cloak of some alternative scientific theory. If a sufficient number of scientific terms are thrown into the mix, it could seem that a stem cell is a stem cell is a stem cell—that adult and ES cells are equally promising. But such a contention is simply not true.

While the mode of argument of the pro-life anti-embryo research campaigners may seem similar to the moral analysis that underlay the Warnock Report and the Human Fertilization and Embryology Act in that both base moral positions on scientific facts, there is a crucial difference. Those who maintain that adult stem cells are just as good or even better than ES cells are arguing against the weight of the scientific evidence. In contrast, the formation of the primitive streak at around the fourteenth day of development is an uncontestable scientific fact. Once one accepted the desirability and morality of IVF, embryo research to understand and improve the procedure was clearly going to be required. A line had to be drawn somewhere—no one wanted to sanction experimentation on foetuses grown in the laboratory—and the fourteenth day became the scientifically based moral line.

Spare embryos

We are thrown back upon the subject of ethics and the question to be answered is this: do the new therapeutic goals that drive ES research change the moral calculus that has been applied to the reproductive goals of IVF? I believe that the answer to this question is no. Certainly, as far as the embryo is concerned, stem cell research presents the same ethical issues already thought through in relation to IVF. The cells can be derived from so-called surplus embryos. These are left over after IVF treatment. When not wanted or needed, they languish in a kind of frozen limbo until they are destroyed. In the United States alone, there are an estimated four hundred thousand stored embryos. Why not, it seems reasonable to ask, gain some good from the situation and, with proper informed consent, use these embryos, destined never to be born, to ameliorate suffering and even save lives—use them to derive ES cells? Because I do not believe embryos to be persons and because surplus embryos will, by definition, not be implanted, I believe that using them to derive ES cells is not only morally defensible, it is morally required.

I don't see this source of embryos as a morally superior alternative to therapeutic cloning, although many people do. Both IVF and therapeutic cloning require creation and sacrifice of embryos that will be destroyed in order to benefit patients—to enable them to have genetically related children in the case of IVF, and to cure presently incurable diseases in the case of therapeutic cloning. The second purpose is at least as laudable as the first. The fact that in the case of therapeutic cloning one knows at the outset that a specific embryo will be destroyed is not, to my mind, a morally significant difference. For all these reasons it seems to me intellectually and morally incoherent to accept IVF and reject therapeutic cloning.

The role and rule of law

Laws and regulations can either impede or even strangle science or create conditions that allow it to flourish. In no area is this more apparent than stem cell research. The chosen source from which ES cells can be derived has both scientific and policy significance. Countries such as Australia and the Netherlands allow the use of spare embryos but have enacted moratoria on therapeutic cloning. Finland also allows research on spare embryos. Germany's Embryo Protection Law bans all embryo research, but since June 2002, the importation of already existing human ES cells has been allowed. Committees in France, Belgium, Ireland, and Canada have been examining the moral and policy issues raised by stem cell research. The regulatory environment in Asia is generally less restrictive than it is in the West. Britain is allowing both methods to be used under a strict regulatory scheme. The story of Britain's engagement of more than twenty years with these issues forms a key part of the scientific, regulatory, and moral story of stem cell research. And it occupies a prominent and well-earned place in this chapter.

In the United States, there remains turmoil and uncertainty. Clonaid's announcement that it had produced a human clone named Eve triggered great alarm and a move to ban both reproductive and therapeutic cloning. American politicians have been taking rather tentative steps into an area that they have long neglected. For almost thirty years, abortion has been considered the third rail of American politics. As with the third rail on the subway tracks, touch it and you are dead. So elected officials have left the entire area of assisted reproduction unregulated at the national level. However, since 1994, there has been a ban on federal funding of embryo research.

Moral gymnastics?

After the publication of the Gearhart and Thomson papers, proponents of research were determined to get around that ban and carve out an exception for embryonic stem cell research. So they came up with a legal sleight of hand. An embryonic stem cell did not have the crucial characteristic of an embryo—it could not, on its own, become a human being. So, according to this legal interpretation, as long as someone not federally funded had derived the cells from the embryo, research using the stem cells themselves could go forward without violating the ban. Germany has embraced the same analysis.

Although pro-life activists protested against allowing any form of ES cell research, this time there was an equally impassioned political counter-force—patients' lobbies, representing all manner of diseases for which stem cell research held out the hope of cure. Nearing the end of his second term, President Clinton supported both the research and the legal interpretation, and guidelines were drafted by the National Institutes of Health (NIH).

When they came into office, the Bush Administration asked for a review of the guidelines. Poll after poll showed that the American people wanted ES cell research to proceed. Influential members of the Senate who had consistently opposed abortion, such as Orrin Hatch, Republican of Utah, made a distinction between a foetus in the womb and an early embryo in a dish. In August 2001, after months of agonizing moral deliberation, President Bush came up with his own plan, a plan that did not need the approval of Congress. He would allow federal funding for the approximately sixty preexisting embryonic stem cell lines that the Office of Science Policy at the NIH had identified. Thus, no additional embryos would be destroyed to derive cell lines, except of course in the private sector where derivation could go on regardless of Bush's scheme. But the public sector effort, while ongoing, has been stifled, if not crippled.

There is another impediment to American research—a broad patent granted to the University of Wisconsin (the academic home of the researcher James Thomson). Since the President's decision, the university foundation that administers the patent has been trying to cooperate with academic scientists, but they found it necessary to sue the Geron Corporation, the company that had funded Thomson's research, for alleged interference with their plan to distribute the cell lines. Although the case has been settled, intellectual property issues and the restrictions on access that they produce still hold the potential to frustrate and discourage research.

There has been widespread and mounting concern

among scientists that the cell lines available for federal funding, now estimated to be less than a dozen, are simply not sufficient for productive research to be possible. Unless the prohibition of federal funding to derive new embryonic stem cell lines is lifted, American scientists will be locked into yesterday's technology in tomorrow's world. For example, all of the cell lines eligible for federal funding were grown on feeder layers of mouse cells—thus rendering them unsuitable for clinical use. In 2002, a team in Singapore led by Ariff Bongso published a paper demonstrating that ES cells could be grown without the mouse component. Since these cell lines were derived after August 2001, American researchers funded by the federal government will not be able to use them.

Therapeutic cloning in the United States has also faced an uncertain future. Just two weeks before President Bush announced his decision on stem cell research funding, the House of Representatives weighed in with respect to one means of deriving these cells—therapeutic cloning

After a brief and remarkably uninformed debate, the House passed a bill that criminalized both reproductive and therapeutic cloning. In order for a law to take effect in the US, it has to be passed by both the House and the Senate and then signed by the President. A very good compromise bill died in the Senate and, with the new Republican majority and Senate majority leader, Bill Frist, a transplant surgeon who opposes all forms of cloning, a ban on therapeutic cloning is at least a possibility. In the new session of Congress, the House passed a total ban yet again. At the United Nations, the American delegation has been pushing for an international ban on all forms of cloning, seeking to give global reach to the views of the current administration. Perhaps stories about the Clonaid project, allegedly mounted in laboratories across the world, will energize advocates for a total ban, but the British—having made a major commitment to therapeutic cloning—would certainly not be willing to sign on to a ban that encompassed more than reproductive cloning. The British, among others, are also fighting against a broad ban on embryonic stem cell research that has been proposed in the European Parliament.

As America sinks further into the morass of scientific and legal uncertainty, Britain can only benefit. Scientists in Britain can derive their own cell lines. They can go on and on deriving them, until they get what they need—what works best. The regulations are clear; the environment is predictable. Scientists don't have to stop off at their lawyer's office on the way to the lab. In September 2002, the Medical Research Council announced the creation of a stem cell bank to enable all researchers, public and private, to have access to the best cell lines available. The commitment to foster this research reaches to the very top of the British government; Prime Minister Tony Blair is a vocal and enthusiastic advocate.

The baroness and the bishop

Few countries have embraced stem cell research as eagerly as has the United Kingdom. None is better prepared by its history to nurture and control this technology. The British discussion of cloning started early and it has been thoughtful, serious, and thorough. And it has been fascinating, suspenseful, and gripping. I was at the House of Lords on 22

January 2001, when the United Kingdom became the first country to bring therapeutic cloning under a regulatory umbrella.

On that cold and windy afternoon, two days after the American presidential election, I walked into the Peers' entrance of Britain's House of Lords. After receiving a warm greeting from the doorkeeper, who, resplendent in white tie and tails, greets the Lords and Ladies as they arrive, I took a seat in a comfortable armchair next to a roaring fire and awaited my hostess, Baroness Warnock. The most famous public philosopher in Britain and charter member of the group of policy makers and other persons of influence who are generally referred to as 'The Great and the Good', she has both a brilliant mind and an impish sense of humour.

As we sat at a table with a magnificent view overlooking

24. The Houses of Parliament

the river, Lady Warnock told me how she had decided to vote on the bill that would come before the House that afternoon. The regulations, if approved, would make Britain the first country in the world to acknowledge as legal and then formally regulate therapeutic cloning. She would, she said, vote against the Alton amendment, a legislative proposal crafted by pro-life opponents of the research, a proposal which was designed to kill any chances of the work going forward for at least two years. She would vote for the competing amendment, tabled somewhat confusingly by Lord Walton, a medical doctor. Neither her colleagues in the House nor the rest of the world would learn of her decision for three more hours. What the Lords had to vote on in January was an expansion of the permissible uses of embryo research under the Human Fertilization and Embryology Act to include developing therapies for serious diseases. There was no provision in the new regulations that would explicitly legalize therapeutic cloning because it was the government position that this way of creating an embryo was already covered under the provisions of the Act that regulated embryo research. For about two months, between 15 November 2001 and 18 January 2002, that interpretation was thrown into doubt and the whole regulatory scheme seemed under grave threat. (More about this a bit later.)

The overwhelming votes—in the House of Commons in December and in the Lords in January—in favour of allowing such research certainly did not come easily. Indeed, the fate of the legislation remained in doubt as afternoon turned to evening and then late evening on 22 January. The future of stem cell research in Britain—and with it, the UK's primacy in this crucial area of biomedicine—rested in the hands of what at first seemed to me to be the most unlikely of institutions, the gloriously anachronistic House of Lords.

I wondered how this venerable and rather stuffy legislative body would handle a discussion of complex cutting-edge science and difficult moral issues. From my perch 'below the bar', two rows behind Margaret Thatcher, I watched forty members of the Lords rise to address their fellow Peers and discourse with passion, wit, and eloquence on the moral and medical questions at the heart of this research. With a couple of Gilbert and Sullivan-like exceptions they evinced a stunning command of stem cell science.

As the hours advanced, an air of excitement and expectancy filled the House. More and more Peers arrived, as did ministers and other officials. People stood in the aisles, listening to the speeches. Little groups formed and soon were deep in earnest conversation at the back of the chamber and then retiring to the hall behind. The tension was palpable. When Lord Walton, a neurologist and the author of the pro-stem cell research amendment hastily cobbled together as an alternative to the Alton amendment, rose to speak, there was utter silence in the chamber. His brilliant, passionate, and riveting speech about the medical promise of stem cell science almost brought the House to its collective feet. The momentum was with him. That was clear.

Another extremely influential vote—in additional to Baroness Warnock's—was very much in question, that of Richard Harries, the Bishop of Oxford, who was later named the Chair of the House of Lords Select Committee on Stem Cell Research. During his eloquent speech to the House, it became apparent that the suspense involving his position would not soon end. He told his colleagues that he was still wrestling with the moral dilemmas surrounding therapeutic cloning. His influence was central on that January day, and it continued to be felt in his role as Chair of the Select Committee. The Committee's report was issued in February 2002, and, coincidentally or not, shortly thereafter the HFEA granted its first licences for ES cell research under the new scheme.

The debate in the Lords, as in society in general, was not just about science and medicine. There was also a metaphysical dimension. The clergy was there in force. The Bishop of St Albans spoke with seriousness leavened with charm and good humour about the difference between knowledge and wisdom. He ended by inviting his fellow Peers who recognized themselves in his description of a truly wise person to make themselves known to him at a later time. It was not until almost 10.30 p.m.—eight hours after the debate had begun—that the Peers left the chamber to vote. They returned for the announcement of the results. The Alton amendment, which would have effectively scuttled the research, was defeated. 'Their Lordships divided: Contents 92; Not-Contents 212.'

The regulatory saga of therapeutic cloning in Britain was far from over. Just four days after the Lords vote, a court challenge to the regulations was brought by a group called the Pro-Life Alliance. Legal proceedings were protracted, and on 15 November 2001, a High Court judge announced his decision. He had been asked to decide whether the definition of an embryo, written into the 1990 Human Fertilization and Embryology Act, could embrace an embryo produced by nuclear transfer. He said no. Thus, no provisions in the Act could be applied to cloned embryos. There was, he found, neither a mechanism for prohibiting reproductive cloning nor a basis for regulating therapeutic cloning.

Meanwhile, given that there was no ban in law on reproductive cloning or any licensing requirement before it could go forward, Dr Severino Antinori, the Italian doctor who has been advocating using reproductive cloning as a treatment for male infertility, threatened to come to Britain to give it a try. He and his American colleague Panos Zavos had prom-ised to initiate the first pregnancies by cloning by the end of 2001. They said they had many people from many nations clamouring for their services and two hundred infertile couples, including several from Britain, who were eager to have the husband cloned and the baby carried to term by the wife.

After the court decision, Antinori seemed to be in a great hurry to clone some British people—on British soil. As unlikely as it was that he would succeed, the mere threat of his arrival galvanized the government into announcing its intention to introduce emergency legislation to ban reproductive cloning. The legislation was passed by both Houses by the end of November. The government also took the adverse High Court decision to the Court of Appeal. The status of therapeutic cloning remained in limbo until 18 January, when the Court of Appeal in effect reversed the judgement of the High Court. Therapeutic cloning would proceed in the United Kingdom and it would do so under a proper regulatory scheme. In a variation on Gertrude Stein's 'A rose is a rose is a rose is a rose', the Court found that an embryo created by nuclear transfer rather than by fertilization was an embryo nonetheless. The Pro-Life Alliance then appealed to the highest court in Britain, the judicial council of the House of Lords which is called the Law Lords. In March 2003, the Lords upheld the ruling of the Court of Appeal.

On the slippery slope—or not?

Slippery slope arguments can be valuable, but they are often the last refuge of those with nothing else to say. 'Where will it end?' is the rallying cry of opponents of the latest scientific advance. In the case of cloning, the argument has been frequently articulated to make the case against therapeutic cloning. Allowing it would, we are told, ineluctably lead us down the path towards reproductive cloning. The basic idea is that if you do *x*, which is morally acceptable, you will end up doing *y*, which is not. You can slide down from *x* to *y* either because there is no way to distinguish between them conceptually or because the existence of *x* creates a social climate that would be receptive to *y*. Neither of these conditions exists in this case where therapeutic cloning is *x* and reproductive cloning is *y*. Therapeutic cloning would be done to develop therapies for sick people—not to replicate them. And the available medical option of cell therapies would be utterly irrelevant to any determination on the permissibility of reproductive cloning.

Opponents of therapeutic cloning allege that there is another kind of slippery slope—a scientific one. Research would produce improvement in the skills necessary to achieve the first part of the process, nuclear transfer and the triggering of development, thus providing knowledge for those who would seek to implant an embryo and produce a child. Opponents also point to purely practical considerations in their arguments supporting a total ban on cloning. How, they wonder, could unscrupulous people be prevented from spiriting an embryo clone out of the lab and implanting it in a woman's uterus? After all, what sanction could the law employ if this indeed did happen? Surely not a forced abortion.

There are two ways to interrupt such a slide. The first is well-drafted legislation that bans reproductive cloning, but allows carefully regulated cloning for therapy. This has been the British solution. The second may seem rather peculiar or even scary—recourse to animal eggs instead of human eggs in the early stages of therapeutic cloning research, but not for actual therapy. Using animal eggs as hosts for human nuclear DNA would provide a brake on the slippery slope towards reproductive cloning simply because any embryo so derived would almost certainly lack the ability to gestate normally and produce a child. In 1998, an American company, ACT, announced that it was experimenting with human nuclei and cow eggs. In September 2003, Panos Zavos announced that he had perfected his human cloning technique by working with cow eggs. He said that he would implant a cloned human embryo (produced with all human material) into a female surrogate the following month. More recently, in spring 2002, the *Wall Street Journal* reported on efforts to carry out therapeutic cloning with cow eggs in Korea and with rabbit eggs in China. The author of the articles, Antonio Regalado, identified the Chinese researcher as Huizen Sheng, a Chinese-born developmental biologist who had spent almost a decade working at the US National Institutes of Health. In February, about three weeks before the China article was published, persistent calls and e-mails from *Wall Street Journal* reporters began to arrive at Professor Sheng's laboratory. I know about this first hand; I was there.

Another American was then in China—a very prominent one—President George W. Bush. Indeed, as I prepared for my first visit to Professor Sheng's lab, I flipped on the television in my hotel room and watched the President give a televised address to the people of China. He exhorted his audience to embrace the virtues of liberty, tolerance, and religious freedom. While the Chinese definitely have many lessons to learn from the West in these three areas, in the stem cell context, the irony was inescapable: American scientists were being deprived of their freedom to mount the prodigious research effort of which the American academic science community is capable—deprived because of the President's strong and sincere religious beliefs. His ideology, albeit deeply felt, was damaging the prospects of US stem cell research. Meanwhile, China was forging ahead.

We went to the lab by taxi, through the busy, teaming streets of Shanghai, streets thronged with people and full of energy and enterprise. The taxi stopped at the door of the busy emergency room at Xinhua Hospital. I was baffled by the site—it seemed so ordinary, even primitive—until we entered an elevator and then climbed a flight of stairs. At the top, a door opened to reveal a completely different world: a cutting-edge lab doing cutting-edge science. Between them, the team leaders at the lab had thirty years' experience in the United States. They had returned home to do their research. The policies that constrain American scientists simply do not apply in China.

These Chinese researchers did not decide to use rabbit

25. The morning commute

eggs for moral reasons; their motivation was strictly practical. Their lab is above an emergency room, not next door to an IVF clinic or near a slaughterhouse. Thus, they had no access to a plentiful supply of either human or cow eggs. So Professor Sheng and her team turned to rabbits, fusing their enucleated eggs with human skin cells (mainly foreskin). In a paper published in Cell Research, a peer-reviewed journal based in Shanghai, they reported that they had cloned more than 400 embryos. About a quarter of these survived to the blastocyst stage—the point at which embryonic stem cells can be derived, and the researchers maintain that they did derive these cells. While many scientist in the West are excited by the work of these Chinese researchers and hopeful that it will produce new insights, they are uncertain that enough evidence has been presented to confirm that these were true ES cells.

Professor Sheng has no intention of implanting any of these embryos; her goal is therapeutic not reproductive cloning. But, as I said earlier, even if implantation were attempted, an embryo produced by combining an animal egg (with its mitochondrial DNA) and human nuclear material would almost certainly not gestate normally; it would therefore be unable to evolve into a human being. Even if it could, however, the small bit of animal DNA that might persist in the mitochondria would certainly not be able to produce large rabbit ears!

People can and do disagree about the morality of cloning for therapies. I believe that the practice is morally acceptable and even morally required. So it qualifies as an x in my discussion of the slippery slope. But what about the morality of y: reproductive cloning? If the slippery slope argument has any validity, what lies at the bottom must be beyond the moral pale. Is it? In the next chapter, we will examine some of the arguments for and against.

26. Sir Walter Raleigh and his son, Walter

5
A Chip off the Old Block: Cloning for Human Reproduction

[T]o claim a right is an essentially public act, a demand for justice, or for what is thought to be due to oneself or to others. . . . In the context of assisted conception, the only right that could reasonably be claimed would be the right to attempt to have a child.

Mary Warnock

It's retrograde to clone . . . There are other ways of making people identical. We can put them through the same schools and subject them to eight hours of tv every day. That works a lot better.

Steen Willadsen

We are going to make an attempt to clone a patient directly into his suit so that he'll be completely dressed at the end of the operation.

Woody Allen, *Sleeper*

I can't be the clone. I'm me. That's gotta be the clone.

Multiplicity

Human shipwrecks

Cloning by nuclear transfer—using the odd skin cell to reconstruct a genetic duplicate of a person—has transformed an old saying about the uncanny likeness of sons to their fathers from metaphor into reality, and a very literal reality indeed! Is human reproductive cloning possible or even imminent? Yes, probably. Animals from ten mammalian species have been cloned and there is no reason to think that humans cannot be as well. It's a matter of when, not a matter of if. Indeed, it may have happened already.

Is this a prospect for celebration, or consternation? Is human reproductive cloning immoral? Well, most people seem to think so. Certainly, it would be immoral if done anytime soon. Macbeth's urgent injunction about a very different matter—'If it were done when 'tis done, then 'twere well it were done quickly'—applied to cloning is a recipe for trouble. It's simply unsafe.

What would it mean to be a cloned baby born in a month's time or a year's time or even two? Given what we know about the incidence of abnormalities in every species that has been cloned (again, not necessarily every animal), it could be a very grim existence indeed. The animal data is sobering; the catalogue of abnormalities extensive; the severity of the problems extreme. While animal data cannot always be translated into accurate predictions about humans, what we know about the lives and early and often painful deaths of many four-legged clones provides ample reason for disquiet.

How could it be morally acceptable to produce a baby in this way and then watch and wait to see when and whether epigenetic havoc ensues? As we saw earlier, cloning has produced lambs that gasped for air, with blood vessels twenty times larger than they should have been. There have been liver, kidney, and immune defects. Some scientists even maintain that every clone is somehow defective, but there is disagreement on that point.

Where there is no disagreement among the experts is this: contrary to the reassurance of proponents and alleged practitioners of human reproductive cloning, there are simply no tests, genetic or otherwise, that could prove that a cloned embryo in the process of gestation or a cloned baby already born is free of defects. Preimplantation genetic diagnosis (PGD) is one of the tests put forward by proponents as useful in preventing what one of them has called 'human shipwrecks'. Although preimplantation genetic diagnosis of embryos has been practised for almost fifteen years, it is useless for uncovering the sorts of defects that the process of cloning seems to produce. PGD can reveal genetic changes or damage (mutations), but not the flaws in gene expression (whether genes are turned off or turned on in order to accomplish a particular task) that are believed to lie at the heart of cloning-induced abnormalities.

For a musical analogy let's turn to the orchestra. To know that a first violinist does not have any broken strings on his instrument is certainly good news, but it's no guarantee that he will play all the notes he is supposed to play and refrain from playing those he is not. And it's no guarantee that he will make his musical entrances and exits at precisely the right time during the concert tomorrow night or, indeed, the concerts for the remainder of the season. As it is with sounding the instruments appropriately, so it is with appropriate gene expression.

A genetic mutation, for cystic fibrosis, for example, would be apparent at the eight-cell stage of embryo development and persist through the life of any resulting child. In contrast, gene expression in a clone could be normal at the embryo stage and abnormal later on in development. In cloned animals that have died, autopsies have revealed liver cells, for example, that were simply unable to perform their functions because they had failed to differentiate properly. The necessary genes were simply not turned on.

Is it morally right for a scientist or physician intent on being a cloning pioneer to take the risk of causing illness, disability, and early death of a human being? The answer, of course, is no. All the reputable cloning scientists that I know believe that it would be grossly unethical to clone a human given present knowledge and experience. It may always be so or it may not. But if safety ceased to be a concern, what then? How should we think about the ethics of cloning in order to decide whether there is anything intrinsically immoral about it?

Before we embark on this analytical journey together, we have a preliminary task: to clear away the moral fog surrounding reproductive cloning. Stop any three or four or five people in the street and ask them about their attitude towards cloning. I have done precisely that. What I have found is that people are invariably troubled, their reactions almost always negative. Why and about what? Clones are seen as dangerous, damaged, and even diabolical. Those who would produce them are seen as irresponsible or worse. Those who would clone themselves are viewed with disdain. Clearly our collective literary and cinematic heritage has contributed to such views, but so has watching the evening news.

THE NATIONAL ENQUIRER
PRESS
The New York Times
BABY CLONE BORN SAY SPACE ALIEN BELIEVERS
MIKE THOMPSON © DETROIT FREE PRESS
"SHEEZ, WHERE DO THEY COME UP WITH THESE STORIES??"

27. Cartoon

There are two other factors at work as well. The bizarre nature of some of the proponents of cloning (the Raelian sect being the most stunning example) has had a pronounced effect on public attitudes. When there is disturbing news about allegedly successful attempts at human reproductive cloning by people who are, at the very least, wildly eccentric in their beliefs, dress, and behaviour, the messenger cannot help but be identified with the message. Tales of secret laboratories in unidentified countries deepen the sense of unease. These so-called cloners seem strange and scary; hence, so do cloning and clones. Rational discourse about the validity of their claims has been difficult, if not impossible. Headlines in the tabloid press have borne a strange resemblance to those in staid and serious newspapers.

Intuitions on morality

Complicating our efforts to examine cloning ethics with some dispassion is the web of beliefs and preconceptions that we all share—a web made up of what I call our moral intuitions. When I speak of moral intuitions, I am referring to a range of attitudes and beliefs, ideals and aspirations, that we use to guide our lives and formulate our responses to our world and to those who inhabit it. By nature, these intuitions are often the sorts of emotionally charged, metaphorically laden, visceral responses that are evoked by the literature and films that I described in Chapter 1.

Going back to our person in the street who says he is morally opposed to cloning, what sort of reasons might he put forward? 'Cloning is unnatural' and 'scientists are playing

God' are two very common reactions. These expressions—so often offered as terms of opprobrium rather than subjects for thoughtful analysis and exploration—do not seem to me to be particularly helpful or enlightening. They almost never serve as a way in to a deeper discussion of the morality of reproductive cloning. On the contrary, they are more commonly advanced as the first and also the last words—to offer a view already firmly held, to preclude a discussion rather than to initiate one.

The word 'natural' can have many meanings, of course, but I want to concentrate on two related ones. The term can refer to an Eden-like existence; it can also be used to describe a sort of purity and integrity that is opposed to what is artificial. The first idea of nature stems from a rather charming, albeit naive belief that we would all be better off in a state of nature devoid of human interference. In Britain, Prince Charles is perhaps the best-known proponent of this point of view—articulated most memorably by him in his Reith Lectures for the BBC.

Nostalgia for an Eden that never was is patently illogical; our ancestral hunter-gatherers were certainly not like the eighteenth-century philosopher Rousseau's noble savages, although they certainly were savage. This sort of thinking often lies behind attacks on biotechnology and the scientists who work in the field; they are seen as mischievously or even malevolently tampering with what should be left alone. This is the criticism particularly and most vociferously made with respect to genetically modified food. The context in which Mary Shelley wrote *Frankenstein* is redolent of such ideas; the monster that Dr Frankenstein assembled, albeit made of flesh and blood, is a product produced in an utterly unnatural manner—without sex. It is the scientist who 'gives birth' to this artificial form of life and it is science that is blamed for the terrible consequences. In general terms and in current usage, the natural versus artificial distinction is not overtly anti-science but it certainly devalues its achievements. And the contention that we should embrace the natural and reject the artificial, if it carried the day, would deprive us of the benefits of antibiotics, vaccines, assisted reproduction techniques such as IVF, and even spectacles.

Although moral intuitions have value in certain areas of our lives, they are not always dependable and they can even be quite immoral. Take racial or religious prejudice, for example. We all have moral intuitions as products of factors as diverse as our parental environment, education, and religious background. They cannot be expunged; but they can and should be educated so that we can distinguish between the good and the bad, the helpful and the useless. Without such education, reliance on moral intuitions as a basis for moral judgements can produce cloudy thinking, immoral results (like bigotry and xenophobia), zealotry, and a rejection of science and its benefits for human health.

IVF decoupled reproduction from the sexual act. Cloning goes one step further; using the technique for reproduction obviates the need for two parents, for the coming together of egg and sperm. There is no shuffling of genes, and if you don't believe in the variability of that shuffle, just take a good look at the siblings in one large family. Cloning for reproduction would thus be quite revolutionary; it would take us out of the paradigm of sexual reproduction into the paradigm of twins. But these would be twins with a difference; one could be born far later than the other, even a generation or more apart. And they would not be twins produced through sex; the new person would be created by design as the twin of an already existing person. Women could dispense with men altogether. Using a skin cell (containing all the instructions needed to reconstruct a new person) and her own egg, mini me could be created by only me. Men would still need a woman's egg and a woman's uterus but not necessarily a woman partner.

As we saw earlier, assisted reproduction technologies have long been seen as 'playing God', as the reproductive equivalent of Prometheus stealing fire from the gods. And, mindful of the hubristic efforts of Prometheus and Frankenstein and their dire results, many people worry that the cloning enterprise will produce horrific and uncontrollable consequences. Those who make this argument—more properly described as those who use this phrase—see our entrance and exit into this world as God's domain. If science and medicine intervene to make things happen that would not and could not have happened without them, science is seen as displacing, even replacing, God.

While IVF gives scientists control of the environment in which fertilization takes place, so-called displacement of God in this context is about method rather than essence. Cloning is different. The essence of reproduction—the coming together of sperm and egg—could be rendered unnecessary. An ordinary body cell could be used to reconstruct another person just like you or me—at least, genetically speaking. For so many reasons, cloning is particularly susceptible to an intuitive approach to morality. There is so much metaphorical baggage. The whole idea seems wrong. It feels odd, even horrible, like something out of *Frankenstein* or *Brave New World*.

So it must be wrong. But a visceral, instinctive shudder does not a moral argument make.

The fertile and infertile progenitor

If we are not going to rely on the safety argument or on our moral intuitions, how then can we fashion an approach and stake out our own moral position on reproductive cloning? Perhaps a good way into our subject is to line it up against another paradigm—assisted reproduction. Then we can ask whether cloning for reproduction is a mere extension of what has been on offer as a treatment for infertility. Is it that or is it something essentially and intrinsically different?

There are three broad categories of cloning to produce a child, categories marked out by the different motivations involved: reproduction, replication, resurrection. That being said, the categories overlap to a considerable degree. All cloning is, strictly speaking, replication as its essence involves duplicating the genome of the progenitor. Cloning to produce a child is reproductive as opposed to the other type of cloning—therapeutic cloning—whatever the motivation might be to create that particular child. I will use the terms in a rather specific way, however: cloning for reproduction has as its goal producing a child; cloning for replication aims to reproduce a particular genome; cloning for the purposes of resurrection (an impossible dream though that surely is) has as its goal the perpetuation of a unique life that has already ended. I will approach these categories in a variety of ways, using some of the traditional tools of ethical analysis: rights, the balancing of harms and benefits (so-called consequentialist ethics), the slippery slope, and a brief excursion into Kant's categorical imperative.

Moral and legal rights to reproduce were invoked long before Dolly the sheep demonstrated that cloning an adult mammal was scientifically possible. For example, American constitutional law recognizes such a right as part of the larger category of procreative autonomy, which also encompasses abortion and contraception. This right is a very special sort—a fundamental right—meaning that the government would have to demonstrate a compelling interest in order to abridge its exercise. Despite the special status of reproductive rights, however, the United States constitution does not confer a duty or obligation on society, through the agency of the government, to provide the means by which individuals can have a baby—or have an abortion. Thus procreative autonomy is termed a negative right (basically freedom from interference) rather than a positive one. Rights, be they negative or positive, are not absolute. Limitations on their exercise are required by the imperatives of our having to live together in society. Our freedom to do as we wish is restricted by the interests of others and our behaviour is restricted by the injunction, articulated by John Stuart Mill, that we must not harm other people.

Some countries recognize a positive right to provide fertility treatment for those who need it. The range of available treatments is influenced by the specific social and medical consensus. At present and for the foreseeable future there would be a consensus in virtually no country in the world in favour of providing people with the means to undertake reproductive human cloning. Quite the contrary. Privately-funded efforts, of course, would be a different matter, provided that no laws forbade the practice. In accord with the public consensus, laws against reproductive cloning have been enacted in most Western countries. These views may change, of course, but there is no sign of that at the moment. Nor should there be because cloning at this point would be dangerous and irresponsible on safety grounds.

Those of us who are able to beget children the old-fashioned way (without recourse to fertility treatment) are left alone by society to just get on with it. But if the assistance of a third party is required, be it for fertility treatment or adoption, government regulates these activities and there arises a societal interest in safeguarding the welfare of any actual or potential child. Assisted reproduction in Britain is a carefully regulated affair and the law protects the best interests of children. Reproductive cloning is illegal in Britain so the option of cloning as a remedy for infertility simply does not arise. Other countries, such as the United States, are far less regulated with respect to infertility treatment and, in the case of the United States, with respect to cloning. Let's assume, for our purposes, our potential patients are living in a regulated environment, in London, in 2020.

We are in the Harley Street waiting room of the noted fertility specialist Dr X. Cloning has become a safe and accepted means of addressing infertility, although it is not widely practised. Two couples are waiting to see the doctor. Couple A has a fertility problem; Mr A has no functioning sperm. Couple B is not a couple at all. It is just Mr B, a middle-aged man with three grown children, who has decided that he wants to clone himself. Perhaps, like the scientist Richard Dawkins, he is just curious to see how his clone would turn out.

Do all of these people have an equivalent moral claim? Should we think about them differently? There are three moral perspectives from which we can approach these questions: the relative need of A and B for treatment; the existence, or lack thereof, of alternatives to cloning; the best interests of any resulting child (the clone in this case).

Couple A require some kind of assistance to have a child. Since it is Mr A who has the problem, procuring sperm from another man and inseminating Mrs A would be a possible solution. Couple A has considered and rejected this option because, they say, they do not like the idea of having a genetic stranger in their family tree. Instead, they want Dr X to clone Mr A, using Mrs A's egg. She would carry the baby, a later-born identical twin of her husband, to term. Should Dr X agree to their request?

There is an obvious downside to using an alternative method—donated sperm. We know that it is common even now for children born after AID (artificial insemination by donor) to want to find their biological father. In Britain and in many other countries, the law continues to preserve donor anonymity because that was the condition under which the donation was made. (There is currently great debate on this question so the laws may well have changed by 2020.) Is it in the best interests of a child to have an unknown biological father—to have a black hole where his genetic inheritance should be?

This is not merely a psychological issue. As the knowledge forthcoming from the Human Genome Project gives rise to tests and then treatments for diseases with a strong genetic component, doesn't a child have both a right and a need to know his father? For the male readers out there, how about you? Would you prefer to be a clone, a later-born identical twin of your father, rather than have your father be a stranger who sold his sperm to make some extra money? Or even an altruist who became your biological father but does not want to know you or love you or share your life? Remember that we are now assuming that reproductive cloning can be done safely. Perhaps the cloning option in such cases is not so bad after all.

What other sorts of harms might cloning visit on this child to be? There are speculative harms—about strange family dynamics, about the possibility of incest or at least incestuous feelings. There are also concerns about a clone coming to believe that he would not have an open future—medically or psychologically. Would a clone, some wonder, watch his father/twin become ill and thereby learn his own unavoidable destiny? Indeed, if this were true, it would be a very grave burden, but it's simply not true. Even identical twins, gestated in the same womb, raised together, and subject to similar environments as children, often do not develop the same diseases. Most diseases result from intricate interactions among genes and between genes, environment, and lifestyle. The biological lack of open future argument is just fallacious. But what about the psychological dimension? Would a clone feel that his destiny was limited by the choices and behaviour of his progenitor? No, I don't think so—at least not any more than might occur with children born the old-fashioned way.

Then there is Mr B, the other would-be patient in Dr X's Harley Street waiting room. Because he is fertile and has no medical need for cloning technology, should we and Dr X enquire closely into his motives and also into his fitness to be a parent? Should we and Dr X be suspicious and reluctant to allow the procedure? Enquiries into motives are rather diffi-

28. Cloning mini me

cult; what might seem reasonable to me might seem quite unreasonable to you.

Let's go back to our assisted reproduction paradigm and think about, say, a busy career woman with a highly responsible and time-consuming job who feels she could not do justice to motherhood at that point in her life. Mindful of the ticking of the biological clock, she decides to freeze her eggs for use later on, when time and career permit her to be a better parent—or when she meets Mr Right. At present, such women can freeze their eggs and use them when their fertility is diminished and the quality of their eggs is not as good. How should we view this? And is it morally different from the case of a woman or man who is fertile and wants to be cloned instead of having a child the old-fashioned way?

I would distinguish between the two cases: the first one might seem selfish but the second seems worrying.

Because he has been fertile and has children, we can infer, perhaps rightly, perhaps not, that Mr B wants to clone himself so that he can produce a child with a particular genome: his genome. There are a multitude of harms, albeit speculative, that could ensue for his young clone. Falling ever short of parental expectations, a child could suffer severe psychological problems. As a clone, he could feel as if he were living an echo of a life, having a shadow existence with options narrowed by the expectations of himself and others. If you learned that you had been cloned to be 'like' someone else, would you feel pressured to become that person? To excel in the ways that person excelled? To be a violinist, a baseball player, a physicist? Even though you had been told that the medical fate of your progenitor would probably not be your fate, could you watch that person sicken and die prematurely without being paralysed by the fear that this would be your destiny as well? On the other hand, we all know twins. Despite having been brought up in the same environment and gestated in the same womb, more alike than any clones could ever be, twins are nonetheless fully distinct individuals. They develop in different ways, make different choices, and die of different diseases.

There are two problems with catalogues of cloning-induced psychological horrors. First, of course, they are purely speculative. Second, difficulties that are very similar to those envisioned by dire cloning scenarios can exist even in the absence of cloning. Children of famous or high-achieving parents often find themselves living in the shadow of fame and living very troubled lives. Still, the fact that Mr B is fertile would lead me to consider cloning in his case to be a morally suspect, if not outright unacceptable, use of the technology.

The door to Dr X's waiting room has just opened and another couple, Mr and Mrs C, have come in. Both of them are infertile, with no functioning egg or sperm. They too have another non-cloning option—using donated gametes. But in their case, they would need donated eggs and donated sperm. They might even require the services of a surrogate mother. In such a case, neither Mr nor Mrs C would be the genetic parent of the baby and Mrs C might not even be the gestational mother. The surrogate who would give birth would be the mother under the law; her husband would be the father. And Mr and Mrs C would have to adopt the child.

On the other hand, if the woman had a functioning uterus but no viable eggs—she could even be past menopause—she could give birth to a baby produced through IVF using a donor egg and donor sperm, so at least she would have some relationship to the child and be the mother under the law. She could also give birth to a clone who is her husband's later-born identical twin. The child would be genetically related to at least one of them. It seems to me that reproductive cloning is an option that should be kept open for couples where both parties are infertile, if and when cloning becomes safe. Clearly, however, the set of circumstances would be rare.

Back in Dr X's increasingly crowded waiting room a new patient appears: Mrs D. Her husband is dead. Shortly before he died, as he lay in a coma, Mrs D asked the doctors to retrieve some of his sperm. She wants to use that sperm to become pregnant. If she is denied that opportunity, she wants to clone herself. For the female readers of this book, what do you think? Would you rather have a progenitor parent (and be a later-born identical twin of your mother) or a posthumous parent whom you had never met or, perhaps one day, even a parent who had never been born at all (using sperm or eggs from aborted foetuses)?

In addition to fanciful scenarios of cloning-induced personal harm, there have been many dire predictions about the social consequences of cloning. For example, some people maintain that adoption of the practice—in the place of reproduction the old-fashioned way—will thwart our normal process of evolution and alter the fabric of society. I don't believe any of this for a minute, principally because there is no reason to think that cloning will ever become the preferred method of having babies. For the most part, people who marry want to have a child together. Their child—a product of the wonderful and surprising shuffling of genetic inheritance.

For the first human clones—unless they are cloaked in a veil of anonymity—life may be difficult indeed. You will remember the world's reaction to Dolly and the behaviour that reaction elicited. Dolly was treated like a freak, a media superstar, and she responded by becoming a rather freakish and highly strung sheep unable, at least in the early stages of her life, to avoid the media spotlight. After the birth of her first lamb, Bonnie, Dolly had a quieter life. But, still a celebrity, she was not allowed to roam free and was kept locked up in a barn with many other sheep. It might well be said that it was her clonal celebrity that killed her, and not just a lung infection.

29. The Dionne quintuplets playing with instruments

For a cautionary example of what life might be like for the first publicly known human clones, we have only to look at the real-world horrors endured by the Dionne quintuplets.

Born in 1934, in a small town in Ontario, these first surviving quintuplets were taken from their parents and cared for by the government. They dressed alike even as adults and, as children, were put on show, at the circus! There would be no biological or psychological imperative to treat a human clone any differently than an ordinary child, for they would be ordinary children albeit produced in an extraordinary way. Any freakishness about cloning would be in the eye of the beholder. But therein lies a problem. In the words of Eliza Doolittle, the heroine of the musical *My Fair Lady,* the difference between a lady and a flower girl lies in how she is treated. We can only hope that the first human clones will have a chance for relatively quiet lives, away from the media spotlight.

Gloomy prognosticators such as Leon Kass, the chairman of President Bush's Council on Bioethics, have written that cloning will lead to the commodification and consequent devaluation of human beings. He said virtually the same thing about IVF twenty-five years ago and his grim vision of the future has not come to pass. In *Brave New World*, Huxley described the manufacture of babies by a combination of cloning and environmental conditioning. Kass and others fear that genetic engineering, an unknown possibility when *Brave New World* was written, will in fact bring about just such a world. If Pandora's box is opened, if we put one foot on the slippery slope (the clichés and metaphors are endless), we will be doomed. No law, no social consensus can prevent a slide into the abyss. I don't believe this either. Science can transform medicine and greatly enhance our health and happiness. Reproductive cloning for infertility may play a small part in our future or it may not. If it does, some of the motives for doing it may appear to society to be more worthy than others. In fact, in highly regulated environments like the United Kingdom, some motives may be disqualifying. But there is nothing about reproductive cloning for infertility that is intrinsically immoral.

Cloning for replication

We have discussed various ways in which reproductive cloning could be a treatment for infertility or for other patients who are, for one reason or another, unable to produce a child without assistance. There are two other reasons that someone might want to clone—to replicate a specific genome of a living person and to replicate the genome of someone who has already died. Why would anyone want to replicate a genome? There are both medical and socio-political motives. We will start with the former and go on to the latter.

About fifteen years ago, an American girl called Annisa Ayala was dying of leukaemia in the United States. All efforts to find a compatible bone marrow donor had failed. Her middle-aged parents were desperate. They decided to try to have another baby. Even if Mrs Ayala could conceive, the chance was only one in four that a new baby could provide compatible bone marrow for their daughter. There was much consternation in bioethical circles as news of this case spread. Some condemned the parents for violating a basic moral rule—Kant's categorical imperative. Any resulting child, they maintained, would be treated as a means to an end that had nothing whatever to do with her welfare. Other observers pointed out that Kant's maxim was not automatically violated by the couple's actions. For what the categorical imperative tells us is not that we may never use another person as a means to an end—we do that all the time—but that we may not use that person SOLELY as a means to an end. As it turned out, the new baby, Marissa, born in 1990, proved to be a perfect match for her sister. A bone marrow transplant was performed when the baby was fourteen months old; it was successful and Annisa's life was saved. Her parents have described Marissa as doubly loved—for herself and also for the gift of life that she gave her sister.

As for Kant's categorical imperative, we should remember that parents have children for all sorts of reasons, many of which are to serve as a means to particular ends. They might want a child who could grow up to run the family farm or business, to support them in old age, to balance their family, and, lest we forget, they might need an heir to secure the line of succession of the British monarchy. Having other motives does not make this sort of thinking wrong; having only those motives does. This is true with respect to conventional reproduction and it is just as true with respect to cloning for reproduction or for replication in a medical context.

For those who ask what would have happened had the baby not been a match, I remain unconvinced that she would have been rejected by her parents. But even if there is a chance they could be right, let us fast-forward this case to 2020 and imagine the alternative that cloning could give the family. Had cloning technology been available, would cloning Annisa—to replicate her genome and secure a 100 per cent chance of a compatible bone marrow donor—have been any more morally problematic? I don't see why it would.

Sceptics raise the spectre of the slippery slope, about which we heard at some length in Chapter 4. Here, however, we are not concerned with whether therapeutic cloning leads inevitably to reproductive cloning but whether, if we allow cloning to produce a baby who could donate bone marrow, in a procedure that is not dangerous for the donor, will we inevitably slide towards morally abhorrent practices such as the clone in the cupboard scenario—using a brain dead or even a sentient human clone as a source for body parts. I don't see a danger here because any clone would have the same legal rights as you and I.

Killing a clone in order to remove his heart, for example,

would be murder, just as killing anyone else would be. Short of this gruesome scenario, producing a cloned child to use one of its kidney's to replace the non-functioning kidney of a progenitor could be prevented by proper legislation. Again, I would say that sacrificing medically and morally acceptable benefits because of fear of misuse is the wrong way to look at biomedicine. If we proscribed every technology, every medical procedure, because they could be misused, we would not

30. Dad's cloned army

only be giving up on the rule of law but we would be giving up on medicine.

A second motivation for cloning to replicate a genome is to perpetuate oneself. It is shocking and amusing to discover how many people raise the spectre of someone like Saddam Hussein gaining access to cloning technology. What would he do with it, they ask ominously? I have heard people say that cloning could be turned into a weapon of mass destruction. I really don't understand this mode of thinking unless it relates to the scenarios of science fiction. If Saddam Hussein or any other tyrant cloned himself, what would he get? The answer is not a brutal dictator but a baby—born in a different time, with different experiences, with a long way to go chronologically speaking until he could be old enough to be a threat to anyone.

Cloning for resurrection

As we have seen, many of the supposed harms that could result from reproductive cloning are based on misconceptions. The same is true for many of its alleged benefits—for example: people who believe that cloning will bring back their dead baby or their dead parent. Of course, it won't; a clone would be an independent person—a new life, not a continuation of an old one. Cloning a dead child would not even accomplish its obvious objective—to assuage the grief of the parents. In fact, because the clone would certainly look a lot like the child who died (being its identical twin after all), a parent watching this new child grow up might have the opposite reaction: grief could be prolonged indefinitely because of the constant reminder of the loss. From the clone's point of view, it would be terribly unfair because he or she would be expected to fill the empty place in the lives of the parents. To be a sort of understudy who is called upon only because the person they really wanted is no longer around.

Cloning cannot help us to achieve serial immortality. When you are dead, you are dead. Your clone, as is true with your twin, would be someone else, a separate and unique person. Let us return for a moment to the Harley Street waiting room of Dr X, the infertility specialist. As you will remember, Mrs D came to see the doctor because her husband had died and she wanted to use sperm, taken from him while he was in a coma, to produce a child. At that point, I raised the possibility of Dr X refusing to do the procedure and Mrs D asking, instead, to be cloned herself. But what if she had not just obtained her husband's sperm but also a bit of his frozen tissue? What if she asked Dr X to clone her husband instead of her? She could give birth to the later-born identical twin of her dead husband.

Under these circumstances, which is the better moral course—allowing her to use the frozen sperm or the frozen tissue? The sperm could be used to produce a child that the couple had wanted to have, a child created by the two parents. Granted, because the father would be dead, circumstances would be less than ideal for the resulting child. But what if Mrs D cloned her dead husband, expecting, even if that expectation were not conscious, that he would somehow be restored to her? She would have to be disappointed in the end, and the child, from whom too much was expected, might pay a high psychological price. Allowing use of the frozen sperm would, it seems to me, be the far better course.

It seems a paradox that cloning, a technique seen by many as a threat to identity and authenticity, could be employed in a futile effort to perpetuate or resurrect a unique life. Personal identity is not equivalent to genetic identity. It is a far more complex amalgam of environment, experience, and biography, and the effects of blind chance. For an exploration of the factors that make us who we are—make us 'unique in all the world' in the words of the fox in Saint-Exupéry's Little Prince—please turn to the next chapter and examine with me the fragility of identity.

31. The process of self-awareness

6

Double Trouble: The Fragility of Identity

If they were taken and placed side by side, nobody, absolutely nobody, would have taken it on himself to say which was the old and which was the new, which was the original and which the copy. . . . He even began to doubt his own identity

Dostoevsky, *The Double*

There will be somebody with my name and she'll cook and clean like crazy but she won't take pictures and she won't be me.

The Stepford Wives

Selves, unlike cells, can never be cloned

Thomas Bouchard, Minnesota Center for Twin and Adoption Research

With a single hair from Mozart's head—someone with the skill and equipment . . . and the women . . . could breed a few hundred infant Mozarts. Find the right home for them and we'd wind up with five or ten adult Mozarts, and a lot more beautiful music in the world.

Ira Levin, *The Boys from Brazil*

Fictional doubles

When we gaze in the mirror, we see ourselves. Or do we? Edward James, the subject of the René Magritte portrait, seems to be looking through the glass at someone else. While the image in the painting is counter-intuitive and thus jarring, looking in the mirror and seeing someone else is what we all do. The conscious you or I is on one side and a mere apparition—a lifeless likeness with no independent existence—is on the other. Through the looking glass is a two-dimensional mimic of our actions and expressions. Mirrors split us into the self and the other. It's our double that we see, a double that cannot cast its own reflection. In myth and legend beings unable to cast a shadow or see themselves in the mirror are seen as diabolical, even soulless.

The concept of the double has fascinated psychologists, including Sigmund Freud and Otto Rank. Freud was particularly drawn to the work of the novelist E. T. A. Hoffmann. In his essay on the uncanny, he describes Hoffmann's work as permeated by doubles in various guises. Some just look alike. Others show signs of mental telepathy. Still others undergo what he calls doubling—a dividing and interchanging of the self. Doubles often appear in literature at the moment of death, perhaps because it is at this time that the soul is said to leave the body. They can even be the instrument of death.

Perhaps most frightening of all the doubles are those whose appearance heralds psychological disintegration. An especially chilling example can be found in Dostoevsky's The Double, a short novel which chronicles the descent into madness of a self-absorbed, repressed civil servant, Mr Golyadkin, referred to as 'Senior' to distinguish him from 'Junior', his 'evil twin'. Junior turns up one day, as if he were a mirror reflection come to life, and becomes Senior's nemesis. In short order, Junior moves into his flat, makes himself comfortable in his office, and ingratiates himself with his boss. In the horrifying final scene, Senior is hustled off in a carriage, presumably to a lunatic asylum, while Junior runs along behind, blowing kisses in farewell.

Although a clone of any existing being could never be its exact double—being born a baby there would always be some difference in age—fictional visions of doubles evoke the quintessential albeit scientifically impossible cloning nightmare. You or I or anyone could be displaced and replaced by a counterfeit version, a shadow self, and no one, perhaps not even we ourselves, would be able to tell the difference. Indeed, Arnold Schwarzenegger's cloning film, *The Sixth Day*, has precisely this story line.

At the height of the cloning anxiety of the 1970s two films were released that conveyed the horror of the outer shell of a person being duplicated by soulless creatures who murdered the human originals and then replaced them. In the 1975 film *The Stepford Wives,* scientists are not the villains; indeed they never appear. The malevolent force is the Association, an organization of successful business executives living with their wives in an American suburb called Stepford. It seems an ordinary sort of place except that all the wives look like Barbie dolls—thin, attractive, meticulously groomed, and utterly devoid of personality. These robotic housewives speak in a mechanistic monotone and can't seem to get enough of cleaning, cooking, and shopping. The men of Stepford have replaced their all-too-human wives with automata. Joanna, a photographer who moves to Stepford with her husband, realizes what is happening but is powerless to avoid her fate; she is strangled by her double.

In the last scene, all the wives meet in the supermarket as they wheel their metal carts and mechanical selves around the aisles. This film is a social satire about the quest to be the perfect wife just as much as it is a horror film. It also captures yet another aspect of our worries about human cloning—the fear that it will be used in an attempt to create perfect people. Indeed, the head of the Association describes his plan to the soon to be replaced Joanna, as 'perfect for us and perfect for you'.

In 1978, the year that saw the creation of the first test-tube baby, publication of David Rorvik's *In His Image,* and the film version of *The Boys from Brazil,* a remake of the Cold War classic, *Invasion of the Body Snatchers,* was released. Instead of the earlier film's rather minimalist sense of totalitarian menace, the new version draws rather explicitly upon Huxley's mass production of humans and also upon the cloning concerns then in full flower. Or in this case full vegetation: aliens from outer space rain down on the planet as seeds and form pods next to people who have made the mistake of falling asleep. The petals enlarge and a double emerges, shrouded in a membrane—to replace the now dead person lying next to it. The pod people look like the originals but they are automata, people whose very lack of emotion and free will are useful for a totalitarian society. Like the Stepford Wives, they evoke both the horror of the humanoid and the horror of diminished or devalued humanity.

Doubles and confusion about identity don't have to be scary, of course. Comedy has exploited the mischievous potential of mistaken identity since the ancient Greeks, specifically in connection with twins. Shakespeare also took his turn, with *Twelfth Night* (about fraternal twins, Viola and Sebastian) and *The Comedy of Errors* (about two sets of twins). And the subject still guarantees a laugh in television comedies. Comedies of mistaken identity are not threatening for one principal reason. We, the audience, and the characters themselves are not confused about who they are. It's those around them who seem to be what they are not. And, of course, all is made right in the end.

Nature's doubles

Unless they are strangers to us, real-world identical twins are much harder to mix up than their literary counterparts. Although they share a genome and thus a single genetic identity, each twin has a separate and unique personal identity. As do we all. This particular combination of personality, intelligence, habits, quirks, talents, and attitudes about the world and the people in it can be imitated by flatterers and impersonators but it cannot be transferred from one person to another by cloning or in any other way.

The influence of genes compared with environment on parameters such as personality, cognitive ability, behaviour, and predisposition towards certain diseases has been measured in studies of identical twins who have been raised together, as well as those who have been separated at birth or shortly thereafter. Thomas Bouchard, the founder of the Minnesota Center for Twin and Adoption Research at the University of Minnesota, has been the most important researcher of twins in general and those separated at birth in particular. Most identical twins are raised together. Because they have

32. Pavarotti and clone

shared the same family environment with respect to their parents and to all the other siblings, it is not always easy to tease out the relative contributions of genes and environment. Twins raised apart present scientists with rarer but much better opportunity. Bouchard and his colleagues have told many enchanting stories about bizarre similarities that recently reunited twins discover in each other. One set, termed 'the giggle sisters', could not stop laughing; another pair had the habit of rotating their necklaces when responding to questions, but not when listening to them; a third set of twins would only enter the ocean backwards and just until the water level reached their knees. Clearly, genetic factors determine more than we think in ways we cannot even begin to imagine.

Heritability is the term used to describe the degree of genetic influence on characteristics that vary across a population, such as personality, cognitive ability, and predisposition to certain diseases. It is expressed as a percentage across a wide population, not as a statistic that is specifically applicable when comparing a person to his or her twin. According to Bouchard's studies, personality traits such as extroversion and neuroticism are about 50 per cent heritable. Intelligence—or IQ—is between 60 per cent and 70 per cent heritable. What this means is that, in any population, between 50 per cent of the variance in personality traits and 60 to 70 per cent of the variance in intelligence is due to genetic factors. Our environment accounts for the rest. Although we frequently hear reports about genes for traits and talents as diverse as perfect pitch and circus performing, this does not mean that genes exert their influence directly. Rather, as it is with many chronic illnesses like heart disease and cancer, genes merely confer predispositions. The twin studies show that there are no personality or behavioural traits that are simply inherited, as a mutation that causes a lethal disease could be, for example.

While it is true that the brains of twins are comparable in size and in the relative proportions of the various lobes, these similarities only extend to the gross structures of the brain. Many of the billions of connections between brain cells are established after birth. In a very real sense clones, twins, and the rest of us grow our own brain. Everything we see and hear, everything we touch and taste, what we learn, what we do—all these factors combine to make us who we are. This is why early childhood experience and education becomes so important. Without a rich and stimulating environment we cannot reach our full potential; our brains become stunted, as do our lives.

The environments and experiences of progenitors and their clones would be substantially more different than those of identical twins raised together. They would have different family environments, even if raised in the same household by the same parents; key aspects such as birth order and parental age would vary. Outside the home, the societal environment would also differ, even if the clone and progenitor were rather close in age. With respect to identical twins raised apart, the family environment would differ but the broad social environment would not, at least not so dramatically. Unlike both types of identical twins, clones and progenitors would have different womb environments; even if the birth mother were the same, the pregnancy and all the hormonal and other events that marked it would be different. Both twins who grow up together and twins who do not would all live during the same generation; they would partake of the same very broad social and cultural atmosphere. Clones and progenitors, on the other hand, might grow up years or even a generation or more apart. If they did, the environmental variation would be enormous.

Designing a dictator

A fantastic albeit fictional cloning experiment is the subject of Ira Levin's novel, *The Boys from Brazil*. While the book sold well, most people who know the story know it from the film. Dr Joseph Mengele, who performed horrific experiments at Auschwitz (many of them on twins), turns up in Paraguay after the war and is up to his old diabolical tricks. Or rather up to some new ones. He has created ninety-four clones of Adolf Hitler, from genetic material obtained shortly before death. In contrast to most of the recent fictional treatments of cloning, *The Boys from Brazil* devotes a great deal of time to scientific detail and especially the relative roles of nature and nurture. Mengele is mindful of the influence of the latter; each child is adopted by a couple chosen specifically to duplicate the environment of Adolf's childhood—the parents' age, their personality, the father's occupation, etc.

Because Hitler's father died when he was 65, when the boy was just 14, assassins are dispatched by Mengele to kill the adoptive fathers. The plot is foiled, however, and at the end we are left to wonder whether any of the boys will grow up to terrorize the world as their progenitor had done. But, of course, they could not. Even allowing for similar family background, the environmental differences between the boys and Hitler remain—where they grow up, when, in what larger societal context, with a very different range of experiences, failures, and successes resulting both from their circumstances and their choices. For we are the product of not just our genes and our environment, but also of what we do and what is done to us. Such a mosaic can never be duplicated.

The contrast between the horrific and the sublime—between Hitler and Mozart—was explicitly referenced in *The Boys from Brazil*. So let us end this chapter with another fantastic experiment: if we had an intact DNA sample from Mozart and could thus clone him, what would this new person be like? Would he be a musical genius or even a musician?

Cloning Mozart

Some families are just more musical than others. The most obvious example is the Bach family—with more than eighty composers, performers, and church music directors spanning a period of three hundred years. As we have seen from the twin studies, personality and intelligence have a genetic component. Musical ability does as well. A study of twins published in 2001 demonstrated that the heritability of pitch perception is extraordinarily high—80 per cent. While being able to pick out the right notes from the wrong ones is obviously not sufficient to make one a musician, it is surely a precondition. Giftedness, and beyond that genius, are far more complicated matters, and the quality of one's environment, especially with respect to genius, plays a very crucial role.

Wolfgang Amadeus Mozart was born in Salzburg on 27 January 1756, the seventh of seven children, five of whom died in infancy or early childhood. His only surviving sibling was his sister, Nannerl, who was five years older than he. Mozart's father, Leopold, was a violinist and composer. His maternal grandfather was also a musician. His sister was extraordinarily gifted at the keyboard but her potential was never to be fully explored. Mozart had two sons. The elder, Carl, tried and failed to be a musician and became instead a civil servant. The younger, Franz Xavier Wolfgang (called Wolfgang), was four months old when his father died. He had the benefit of a good musical education. Although it is said that his mother cautioned him that no son of Mozart could afford to be mediocre, he went on to have an undistinguished musical career.

There are three aspects of Mozart's environment that are relevant to the degree of similarity between him and our hypothetical Mozart clone: the character of the society (the eighteenth century was clearly very different from the twenty-first), the family environment, and the prevalent musical culture. We will examine only two—his family and the character of the musical world in which Wolfgang found himself. When she was eight, Nannerl began keyboard lessons with her father. Her brother, Wolfgang—aged three-and-half—demanded to be taught as well. And he was. (If he hadn't had a surviving sibling just the right age for music lessons, this would not have happened.) By the time the boy was four, the dazzled Leopold was keeping records of his extraordinary musical prowess. Soon, the father, who had been a prolific composer and pedagogue, gave up his own musical career to supervise the education of his children.

By 1762, both children were so accomplished musically that Leopold took them on the road. Just one year later, Wolfgang's and Nannerl's musical fates were sealed: for the next fourteen years, Leopold would travel Europe with only one of his children, his son. Every trip had a purpose and the boy's time was tightly structured. For example, when Wolfgang was thirteen, Leopold took him to Italy to develop his compositional skill and to immerse him in the Italian language. On all of these trips, Wolfgang met composers, practised, gave concerts, and was tutored by his father in subjects as diverse as religion, politics, art, and history. The life of the Mozart family revolved around his potential and his needs, with

33. Mozart with his father and sister

Leopold designing and controlling every aspect of his son's life for the sake of his genius. Because that genius found expression in music, it is to the prevailing musical culture that we must turn next.

The Baroque style was dying. Mozart found a musical world waiting to be remade, redefined. With his extraordinary intelligence, excellent education, wide experience of music, and fertile imagination he was in a position to do just that. The sonata form was being developed by him and others, but I want to concentrate on only one sphere of music—the opera. *Opera seria* was the province of Handel. While its music could be wonderful, the arias were written to express one fairly abstract 'affect' or emotional state, such as love or rage. The action of the drama stopped as they were sung. The arias seemed interchangeable from opera to opera. In fact, singers often did just that. The comic opera of the time, *opera buffa,* had flat characters and stock situations, commonly the pretty young servant girl making a fool out of her rich, elderly, and besotted, or at least lecherous, master.

It is no exaggeration to say that Mozart remade the opera. Serpina (the servant girl in Pergolesi's *La serva padrona*) became Susanna, the wise, loving, and psychologically complex lady's maid in *Le nozze di Figaro*. His characters are fully rounded human beings and the arias they sing are theirs alone. Substitution is simply unimaginable. Ensembles and arias propel the plot forward. There is no resemblance between what Mozart found and what he left us. How could any modern composer, cloned or not, who has heard Verdi, Puccini, and Wagner put himself back into the musical universe where Mozart lived? It's just not possible. Nor is it possible to re-create his family situation; even cloning

34. **The Salzburg marionettes in Mozart's Così fan tutte**

Leopold, and waiting another twenty-five years until he could raise his son's clone, would not do that.

So what would Mozart's clone born in, say, 2005 be like? Wolfgang II would certainly look like Mozart; he would have his features, albeit with a different hairstyle. He would be taller and stockier; nutrition is better now and prenatal care is widely available. Indeed, Mozart may not have been born at all had more than one of his older siblings survived. We would expect Wolfgang II to be highly intelligent and similar in personality to his progenitor because these traits are highly heritable. He would have a higher than average musical ability; he'd probably be quite gifted. But would he be a child prodigy or a musical genius? Well, without the influence of a father like Leopold, he would not be a prodigy, but perhaps he might be a genius. Twin studies cannot help us here; there are simply too few geniuses to study.

Even if Wolfgang II composed music, he would not compose in the style of Mozart. He would not sit down and write Symphony No. 42. The musical culture as it was then is gone. The challenges that Mozart faced and mastered are gone as well. Perhaps Wolfgang II would be a rock musician. It is impossible to say. What we can say with certainty is that Mozart was a product of a highly complex and unique set of circumstances and influences of which his genetic inheritance was but one. It would be impossible to produce a newly minted copy of an artistic genius. But what about copies of the works themselves—of paintings and sculpture, for example? To explore the subject of copies and what they have to do with our ideas about human cloning, please turn to the next and final chapter.

35. Aficionados

Conclusion: There's Only One *Mona Lisa*

> *I am uncomfortable with copying people because that would involve not treating them as individuals.*
>
> Ian Wilmut

> *A twin would be a surprise, but a number, any number, would be a shock.*
>
> Carol Churchill, *A Number*

> *First of all, I am an individual.*
>
> Henrik Ibsen, *A Doll's House*

If I had a reproduction of Leonardo da Vinci's *Mona Lisa* hanging over my sofa, I doubt very much that there would be a queue of people snaking round the corner, all waiting to have a look. Similarly, the copy of Michelangelo's *David,* standing in the Piazza della Signoria in Florence, doesn't draw a crowd of awed observers. No, their destination is the Accademia just up the road. The reason for all the indifference to the statue in the square is obvious; with rare exceptions (such as Roman copies of Greek statues or works created to be in cathedrals or to stand outdoors), museums house originals—the real thing. Everything else is merely a copy, an imitation, a fake—a substitute with which we simply make do when the real thing is unavailable. A copy of a work of art is of lesser value both artistically and financially than the original. And whatever value it does possess lies not in its intrinsic character and characteristics but in the degree of its conformity to the prototype.

As we know, however, a clone would not and, indeed could not, be a mere copy of its progenitor. Unfortunately, however, the idea of clones as copies has found a prominent place in our cloning conversations. Ian Wilmut, who headed the team that cloned Dolly the sheep, used the term early and often as he sought to demonstrate his revulsion at the idea of human reproductive cloning—at what he described as 'copying people'.

But people cannot be copied. Our mirror image will never come to life on our side of the looking glass. Reconstructing a new person using a single cell taken from a progenitor would duplicate only the genome. The clone would be a later-born identical twin. While the copying metaphor may lend itself more readily to clones than twins because, sequentially, the clone follows the progenitor, the term is as misleading applied to one as it is to the other.

Perhaps the most profound and pernicious of the misunderstandings about cloning is that genetic identity is equivalent to personal identity. It is not. Even so-called Siamese twins (more properly called conjoined), who share the same lower body along with every aspect of their external environment, have distinct personalities.

Conceptualizing cloning as copying inspires hopes and fears with no basis in fact. One such hope is that we can use biology to confer a kind of serial immortality on those who will one day die (as we all will) or on someone who has already died. The heartbroken and the bereaved are easy prey for would-be cloning entrepreneurs. Even false beliefs can have morally repugnant consequences. A clone created with resurrection in mind would be valued not for himself, but only for a high degree of conformity with his progenitor—conformity in personality, behaviours, talents, and perhaps even genius. He would lead a life in the shadow—forever compared, forever falling short.

In contrast, if human reproductive cloning could be done safely (and that is a big if), it could serve as a morally acceptable remedy for infertility, especially if both partners were affected. Similarly, cloning to replicate the genome of a sick child in order to produce a sibling with compatible bone marrow could be a morally legitimate option as long as the new child would be loved and cherished, and not viewed as merely a means to an end. Unfortunately, however, it is fiendishly difficult for the law to make distinctions based on motive.

As with babies created by IVF, cloning to produce a child for infertile parents would not result in diminished personhood or humanity. The mode of creation is morally irrelevant, as it is with respect to children conceived the old-fashioned way. Whether the sexual act arises out of love or out of lust, whatever the circumstances and the motives surrounding procreation, the child will be a fully human person, with full legal rights. There can be no clone in the cupboard to be raided for body parts when the need arises.

While making standardized copies of a unique work of art can turn art into a commodity, parenthood achieved through cloning does not make commodities out of humans. But what if someone produced multiple clones of the same person and not just one? Why does this prospect seem so disturbing? Not because it would result in many people sharing one personal identity; it would not. Multiple clones are upsetting because they would not be viewed as children. Instead, they would be more like products, and their creators not really parents. How would one of these clones feel about being (in the words of playwright Carol Churchill) 'a number'? Devalued, would be my guess.

Images of mass production of human clones evoke Huxley's Bokanovsky process that produced dehumanized, subnormal clones to perform tasks for which they were designed to be suited. While this vision is and will remain fantasy, it is nonetheless deeply threatening to our sense of individuality and autonomy—crucial values, especially in Western societies.

Perhaps some of our unease about cloning—about becoming a devalued copy or the template for one or twenty or one hundred copies—has more real-world roots. So many of the stereotypical associations that surround cloning (loss of individuality, control, uniqueness, and our essential humanity) seem all too applicable to the lives we lead. Especially for those of us who live in large cities, work for big companies, and dress very much the way other people do. We seem anonymous, in danger of not being appreciated for our uniqueness, in danger of not even being noticed. Many of our encounters are with people we don't know and will never know; these strangers certainly don't see us as individuals, as special. The fear of cloning resonates so powerfully because of our suspicions that, metaphorically speaking, we already are clones.

We all long to be one of a kind, with the ability to make our own choices, forge our own destiny, write our own autobiography. Would being a clone give us someone else's story to tell or retell, especially if, as the cliché has it, the book of life is written in our genes? No. A clone would have the same potential for an open future and an unencumbered ability to be an independent moral agent. We are, to use another cliché, far more than the sum of our genes. Imagine for a moment that the most crucial, transforming experience in your life had not happened. You did not lose a parent when you were a child. You did not go through a painful divorce. You did not meet a person on a train or aeroplane who somehow gave you a vision of a different path in life—to go to law school or medical school, to become a missionary or aid worker. You never heard that political speech that thrilled you and sent you straight off to campaign for a candidate. Imagine that whatever it was did not happen. Is there any doubt that you would have had a very, very different life?

Of course, we are profoundly affected by our genetic inheritance but, by and large, genes are about predisposition, not predestination. There is only one Mona Lisa. And there will never EVER be another you.

Further Reading

Beddington, Rosa, *Cloning*, Mill Hill Essay, National Institute of Medical Research (www.nimr.mrc.ac.uk/millhillessays/1997/cloning.htm)

Blackman, Malorie, *Pig Heart Boy* (London: Doubleday, 1997).

Bouchard, T. J., Jr., et al., 'Sources of Human Psychological Differences: The Minnesota Study of Twins Reared Apart', *Science*, 250 (1990): 223–8.

Churchill, Caryl, *A Number* (London: Nick Hern Books, 2002).

Cooper, David, and Robert Lanza, *Xeno: The Promise of Transplanting Animal Organs into Humans* (Oxford: Oxford University Press, 2000).

Deary, Ian J., *Intelligence: A Very Short Introduction* (Oxford: Oxford University Press, 2001).

Dostoevsky, Fydor, *Notes from the Underground; The Double*, trans. Jessie Coulson (Harmondsworth: Penguin, 1972; The Double 1st published 1846).

Gregory, Richard, *Mirrors in Mind* (New York: W. H. Freeman, 1997).

Gutman, Robert W., *Mozart: A Cultural Biography* (London: Pimlico, 2001).

Haldane, J. B. S., *Daedalus or Science and the Future* (London: Kegan Paul, 1924).

Halliwell, Ruth, *The Mozart Family: Four Lives in a Social Context* (Oxford: Oxford University Press, 1998).

Hoffmann, E. T. A., *Tales of Hoffmann,* trans. R. J. Hollingdale (Harmondsworth: Penguin, 1982).

House of Lords Select Committee on Stem Cell Research, *Report* (London. HMSO, 2002).

Huxley, Aldous, *Brave New World* (London: Harper Collins, 1994; 1st published London: Chatto & Windus, 1932).

Huxley, Julian, *Essays in Popular Science* (Harmondsworth: Penguin, 1937; 1st published 1926).

Kass, Leon, 'Why We Should Ban Human Cloning Now: Preventing a Brave New World', *New Republic,* 21 May 2001.

—— 'The Wisdom of Repugnance: Why We Should Ban the Cloning of Humans', *New Republic,* 2 June 1997.

—— 'Making Babies—The New Biology and the "Old" Morality', *Public Interest,* 26 (Winter 1972).

Klotzko, Arlene Judith, ed., *The Cloning Sourcebook,* 2nd edn (Oxford: Oxford University Press, 2003).

Levin, Ira, *The Boys from Brazil* (Harmondsworth: Penguin, 1976).

Mazlich, Bruce, *The Fourth Discontinuity: The Co-evolution of Humans and Machines* (New Haven and London: Yale University Press, 1993).

Melchior-Bonnet, Sabine, *The Mirror: A History,* trans. Katharine H. Jewett (London: Routledge, 2001).

Miller, Jonathan, *On Reflection* (London: National Gallery Publications, distributed by Yale University Press, 1998).

Noonan, Harold W., *Personal Identity* (London: Routledge, 1989).

O'Mahoney, Marie, *Cyborg: The Man-Machine* (London: Thames & Hudson, 2002).

Plomin, Robert, and John C. Defries, Gerald E. McClearn, and Michael Rutter, *Behavioral Genetics,* 3rd edn (New York: W. H. Freeman, 1997).

Rorvik, David, *In His Image: The Cloning of Man* (New York: J. B. Lippincott, 1978).

Russell, Bertrand, *Icarus or the Future of Science* (London: Kegan Paul, 1924).

Segal, Nancy L., *Entwined Lives: Twins and What They Tell Us about Human Behaviour* (London: Penguin, 2002).

Shakespeare, William, *The Comedy of Errors* (Harmondsworth: Penguin, 1972).

Shelley, Mary, *Frankenstein* (Harmondsworth: Penguin Classics, 1994; 1st published 1818).

Turney, John, *Frankenstein's Footsteps: Science, Genetics and Popular Culture* (London: Yale University Press, 1998).

Warnock, Mary, *Making Babies: Is There a Right to Have Children?* (Oxford: Oxford University Press, 2002).

Wilmut, Ian, Keith Campbell, and Colin Tudge, *The Second Creation* (London: Headline, 2000).

Wolpert, Lewis, and Rosa Beddington, Jeremy Brockes, Thomas Jessell, Peter Lawrence, and Elliot Meyerowitz, *Principles of Development* (Oxford: Oxford University Press, 1998).

Index